DEVELOPMENTAL AND CELL BIOLOGY SERIES

EDITORS

M. ABERCROMBIE   D. R. NEWTH

J. G. TORREY

THE PROCESS OF
SPERMATOGENESIS IN ANIMALS

THE PROCESS OF
SPERMATOGENESIS IN ANIMALS

EDWARD C. ROOSEN-RUNGE

Professor, Department of Biological Structure
University of Washington School of Medicine

CAMBRIDGE UNIVERSITY PRESS

CAMBRIDGE

LONDON · NEW YORK · MELBOURNE

Published by the Syndics of the Cambridge University Press
The Pitt Building, Trumpington Street, Cambridge CB2 1RP
Bentley House, 200 Euston Road, London NW1 2DB
32 East 57th Street, New York, NY 10022, USA
296 Beaconsfield Parade, Middle Park, Melbourne 3206, Australia

First published 1977

Library of Congress cataloguing in Publication Data
Roosen-Runge, Edward C   1910–
The process of spermatogenesis in animals.
(Developmental and cell biology series)
Includes bibliographical references and index.
1. Spermatogenesis in animals.   I. Title.
[DNLM: 1. Spermatogenesis.   2. Animals–Physiology. QL966 R781p]
QL966.R66   591.1'6'62   76–9169
ISBN 0 521 21233 2
Printed in the United States of America
Composed in Great Britain at the University Printing House, Cambridge
Printed and bound by The Murray Printing
Company, Forge Village, Massachusetts

Contents

Plates 1–14 are between pp. 132 and 133.

Preface

Plans for this book were laid ten years ago. The inclusion of current literature ended essentially by September 1975. The subject of the book is extremely broad. It not only purports coverage of all metazoan animals, but also sets no limits on the aspects of spermatogenesis to be treated. However, the reader should be aware that in both respects there are severe restrictions. Coverage is limited first of all because only in a minute fraction, probably less than a thousandth, of all animal species has the process of spermatogenesis been investigated at all, and from these many were eliminated because results were not sufficiently well documented for comparative considerations. In addition, the literature is so scattered in the whole field of zoology, with all its taxonomic specializations under the headings of anatomy and physiology, embryology and cytology, fertility and reproduction, that complete coverage is out of reach.

The breadth of the discussion of the available data is further limited by conscious emphasis on certain aspects of the subject to the relative neglect of others. I have concentrated on presenting the process of spermatogenesis as the development of a very peculiar population of cells within the setting of the metazoan body. Review chapters and discussion chapters emphasize first of all the constitution of the germ cell population, the interrelationships between germ cells and the associations of germ cells and soma. The morphological differentiation of individual germ and supporting cells is less fully treated, and the endocrine and genetic controls of the process are regarded as an important but supplementary subject. Mammalian spermatogenesis is described relatively briefly, because it has been well reviewed by others and in the context of this book should take its place in the perspective of the whole animal kingdom.

The difficulty of a terminology which has never been standardized for metazoa in general, I have tried to ameliorate by a Glossary and by proposing and using a somewhat unified terminology in the discussion chapters. However, in the review chapters I have used predominantly the terminology of the authors quoted. Problems of taxonomy, about which I know very little, I have circumvented by using one single standard, the second edition of Lord Rothschild's book *A Classification of Living Animals* (Longmans, Green & Co., London, 1965). Where Rothschild indicates alternative classifications I have used Dawes' alternative for Platyhelminthes and Rothschild's (not Hyman's) for Aschelminthes. All of this does not help with outdated nomenclature in the older literature with which in general I have been unable to cope.

In the long time since I made the first plans for the book, I have had much technical support, assistance, advice and encouragement from colleagues, assistants and students who have sometimes caused me to continue when I have been tempted to give up. I can only mention those who have played a major role but I am grateful to all of them.

First of all, Dr Lawrence Picken became interested in my idea and arranged for a stay at the Department of Zoology of the University of Cambridge for seven months in 1965, during which time I was supported by a Special Senior Fellowship of the NIH. Without Dr Picken's advice and encouragement the book would not have taken shape. I am deeply grateful also to those colleagues and friends who at various times provided opportunities for me to spend time in writing and using the libraries on their homeground. These are Professor Thaddeus Mann of Cambridge whose patience, kindness and expert knowledge have been a continuous stimulus, Professor Gerhard Petry of the University of Marburg, Federal Republic of Germany, and Drs Robert Fernald and Dennis Willows of the Friday Harbor Laboratories of the University of Washington. Others who deserve my great gratitude for help in various forms are: Dr Newton B. Everett, Dr Edward M. Eddy who contributed valuable suggestions to Chapters 9, 10, 11 and 12, and Mr Richard V. Clark of the Department of Biological Structure at the University of Washington. Professor Vincent Wigglesworth of the Zoology Department of the University of Cambridge, Dr Daniel Szöllösi whose contribution to Chapter 4 deserves special mention, and Dr R. Billard of the Station de Recherches de Physiologie Animale, France. Thanks are due to Patricia Veno for the drawings of Figs. 2, 3 and 5, to Janet Hannan for Figs. 1 and 4, Virginia Brooks for Fig. 42, Dr B. Gupta of Cambridge University for the electron micrographs reproduced on Plate 8, and Dr R. E. Watermann of the University of New Mexico for the scanning electron micrographs on the cover picture and on Plates 11 and 12. I thank Doris Ringer for editorial work on early versions of several chapters. My special gratitude goes to Janet Hannan who decided during her sophomore year in undergraduate study to help me by typing and editing a scientific book as a part time occupation, and who made a superb job of it. Finally, and importantly, I wish to thank the Cambridge University Press for their competent, considerate assistance.

March 1976   EDWARD ROOSEN-RUNGE

1

Introduction

In this chapter a general introduction to the phenomena of spermatogenesis will be presented. This will be amplified and expanded in Chapters 3 to 8, in which many phyla are reviewed and specific examples of well investigated cases are described in detail. Finally, several comparative aspects of the spermatogenic process will be discussed in Chapters 9 to 12.

A basic outline of spermatogenesis in animals would be easier to compose if a large number of animals had been carefully investigated in this regard. But it is one of the inherent shortcomings and recurrent problems of this book, that satisfying descriptions of qualitative and quantitative detail are available only for a minute proportion of the extant species. The picture of gamete development which the general biologist carries in his mind is the result of accidents of his own education superposed on the vagaries, rather than the systematic progress, of the history of research, and characteristically depends largely on the selection of prototypes of which the best known is that of the mammal. Spermatogenesis in mammals, and particularly in the rat, is the most thoroughly investigated case, and is therefore, most extensively presented in textbooks of histology through which not only medical and paramedical students but also many general biologists receive part of their basic training. In contrast, textbooks of general biology or zoology usually present entirely unspecific, schematic views of spermatogenesis which are so general that they lack accuracy and interest. In addition, they often are based on data from a time 50 or more years ago when information was even scantier than it is now, and their validity is therefore not beyond doubt.

The following general outline of spermatogenesis is not based on a specific animal or phylum, but on a conceptual fusion of features which have been reported consistently and, therefore, today seem important and basic. Obviously this didactic conception is subjective and must be read with the mental reservation that one may want to revise and amend it after having read the whole book.

The word *spermatogenesis* means to most biologists simply the development of a spermatozoon. This is only natural. Since the time of the promulgation of the cell theory when Koelliker first established the derivation of sperm from cells of animals testis (see Chapter 2), a narrow cytological view of spermatogenesis as the development of individual male germ cells has prevailed until very recently. However, it is now becoming clear that a broader view, more dynamic and more "ecological", is entirely possible and, in fact, justified. Fig. 1 shows a diagrammatic representation of the old and the new concept. The concept of spermato-

Two ways of looking at "spermatogenesis"

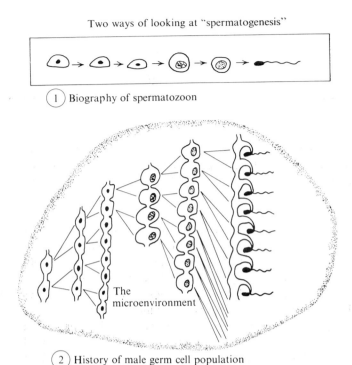

(1) Biography of spermatozoon

(2) History of male germ cell population

Fig. 1. Historical and schematic view of spermatogenesis.

genesis as the biography of the spermatozoon is linear. It leads in simply named stages, i.e. spermatogonium, spermatocyte and spermatid, to the finished highly differentiated sperm cell. In this view, external influences impinging on this developing cell are worthy of consideration but are not indispensable parts of spermatogenesis.

In the new view, spermatogenesis is the life history of a cell population inseparably bound to and conditioned by its environment, the body of a specific animal. As it turns out, male germ cells *never* develop alone. They always grow in close aggregates, usually in syncytially connected clones, so that germ cell influences and is influenced by germ cell. In most animals certain somatic cells take part in this process, "supporting", "nutrient", "sustentacular", "nurse", "cyst" or "Sertoli" cells. The germ cells or the aggregates of germ and somatic cells are set apart from the soma at an early time in development by a boundary layer of somatic cells which serves as a barrier and clearly demarcates an *immediate* environment, e.g. a "cyst" in insects or a "seminiferous tubule" in mammals, from a *wider* environment, the gonad or the body which houses the gonad. The immediate environment not only contains and conditions the process of spermatogenesis but it is inseparable from it. A comprehensive view of spermatogenesis

TABLE 1. *Stages of the development of male gametes*

Term	Name of cells	Characteristics
1 Primordial development	Primordial germ cells	Not sex-determined, extragonadal, mitotic
2 'Pre-spermatogenesis'	Gonocytes, 'stem cells'	Sex-determined, gonadal, mitotic, some cellular differentiation
3 Spermatogenesis		
(*a*) Spermatocytogenesis, mitotic, spermatogonial stage	Spermatogonia	Mitotic, clone formation, cellular differentiation prominent
(*b*) Spermatocytic or meiotic stage	Spermatocytes I and II	Postmitotic meiotic, including two maturation divisions and often cell growth
(*c*) Spermatid stage or spermiogenesis	Spermatids	Postmeiotic, haploid, no multiplication, highly specific differentiation into spermatozoa

in animals must, therefore, include the specific microenvironments of the germ cells. It must also embrace the control mechanisms which serve the communication between soma and gametes, enabling spermatogenesis to proceed in an enormous variety of internal milieus produced through the evolution of organisms.

Before this broad view is taken, however, it is necessary to review briefly the development of germ cells and specifically of male germ cells and establish the pertinent terminology. Germ cells can be identified in many animals long before they become enclosed in the sex-specific gonads. The 'germ plasm hypothesis' which states that there is a continuity in substance ('Keimplasma') from the germ cells of the parents to those of the offspring (Weismann, 1892; Wilson, 1925) has in recent years been tested experimentally and morphologically and evidence for it has become steadily more tangible. The electron microscope has revealed that 'germ plasm' may indeed be a discrete structure unique to germ cells and occurring in similar form and distribution in many vertebrates and invertebrates, although it is too early to maintain that it is present in all animals (for review see Eddy, 1975). The material, often called 'nuage', consists of electron-dense, fibrous accumulations of the order of magnitude of 1 μm or less (Plate 2*b*). In a few cases it has been traced through a major portion of the life of germ cells. Mahowald (1971) followed it in *Drosophila* from primordial germ cells into stages of oogonia and oocytes, and Eddy (1974) identified it in the rat in the primordial germ cells in the hindgut and in spermatogonia, spermatocytes and spermatids in the mature testis. At present the evidence appears better than ever that there is indeed continuity of a specific germ plasm.

The development of male germ cells can be divided into three major stages which are summarized in Table 1. Only the last of these is spermatogenesis proper. The first is the stage of the so-called 'primordial germ cells'. In almost all cases these become first recognizable in an embryo in extragonadal situations, sometimes as in *Ascaris* (Boveri, 1899) in the two-cell stage, sometimes much later as in the yolk sac, of the human embryo. In this first, extragonadal stage the primordial germ cells of both sexes appear to be alike in morphology (with the exception of the chromosome set) and in behavior. They are capable of mitotic division and often of ameboid movement which may, e.g. in mammals, assist them in reaching their gonadal location. The sexes begin to diverge in the second stage with the incorporation of the germ cells into the specific setting of the gonads.

The second stage of development occurs within the gonad. It is relatively easily defined in mammals and often called 'pre-spermatogenesis' (Hilscher & Makoski, 1968). It is the stage in which direct interaction between soma and germ cells appears to begin. In man it lasts more than 10 years from the establishment of the gonad in the second month of pregnancy to puberty. During this phase the germ cells, now best called 'gonocytes' (see Glossary), have certain characteristic periods of mitotic activity interspersed with periods of cell degeneration (Roosen-Runge & Leik, 1968; Hilscher *et al.*, 1974) and there may be subtle morphological changes in the gonocytes. The immediate environment of the gonocytes may change greatly through multiplication of supporting cells and (for instance in man) through a tremendous, temporary development of androgen-producing cells in the vicinity of the germ cells. However, this phase is not always well defined and probably does not exist in the many animals in which the process of spermatogenesis begins almost immediately after the germ cells become established in the gonad. Even in animals in which a pre-spermatogenic phase occurs it is often difficult to define precisely where it ends and where the last stage, spermatogenesis proper, begins. However, in a specific case it may be important to know when exactly the process begins which leads irreversibly to meiosis and the differentiation of the spermatozoon. It is quite clear that during the events of their early, extragonadal and later intragonadal, "pre-spermatogenic" life the germ cells attain a certain stage of differentiation which primes them for the entrance into spermatogonial mitoses, meiosis, and subsequent stages of differentiation. For the rat this story has been told in detail, and it is known that the first spermatogonia are ready for the beginning of spermatogenesis proper on the fifth day after birth (e.g. Roosen-Runge & Leik, 1968; Hilscher *et al.*, 1974).

Spermatogenesis, in contrast to oogenesis, usually proceeds without conspicuous pauses, although in animals with seasonal cycles, e.g. in frogs, there are often long "rest" periods particularly in the spermatid stage. The oocyte often, for instance in mammals, but not in all animals, stops in the dictyate state of meiosis for weeks or years and then begins its final maturation stage under the influence of specific stimuli. In the development of male germ cells pauses, if they do occur, usually take place before proper spermatogenesis begins, in "pre-spermatogenic" stages. Once a certain activation of the gonocyte has occurred,

TABLE 2. *Comparative terminology of spermatogonia*

General term	Term used in mammals	Term used in insects (Hannah-Alava, 1965)
1 'Prospermatogonia'	M, T_1, T_2 (Hilscher *et al.*, 1974)	Predefinitive spermatogonia
2 'Stem cell'	A_s (A_{is}; Huckins, 1971), A_0*	Predefinitive spermatogonia
3 'Proliferating spermatogonia'	A_{pr}, A_{al} (Huckins, 1971; Oakberg, 1971)	Indefinitive spermatogonia
4 'Differentiating spermatogonia'	A_1–A_n, In, B (Leblond & Clermont, 1952)	Definitive spermatogonia
(a) Early stages	'Poussérieuses' (Regaud, 1901), A_1–A_n	'Primary' spermatogonia
(b) Late stages	'Crôutelleuses' (Regaud, 1901), B (In some cases a distinct intermediate class, In, can be distinguished between A_n and B)	'Secondary' spermatogonia

* A_0 is a specific term for a 'reserve stem cell' (Clermont & Bustos-Obregón, 1968), and not entirely synonymous with the term 'stem cell'.

development proceeds through spermatogenesis in a series of fairly well known steps.

The first stage of spermatogenesis proper may be called spermatocytogenesis, the "multiplication stage", and the cells ready for or in this stage are called *spermatogonia* (for origin and further explanations of this and many following terms, see Glossary). It begins with division of cells often morphologically identifiable as 'primary spermatogonia'. These are usually relatively large and often contain chromatin in dusty dispersion with one or two prominent nucleoli. It appears that a primary spermatogonium must divide at least once, and usually a number of further mitotic divisions must follow before the resulting cells are ready to enter into the meiotic phase. In general, the daughter cells of primary spermatogonia and their progeny are called 'secondary spermatogonia'. The number of successive spermatogonial mitoses is species-specific within narrow limits, usually three to five (Table 8, Chapter 9). Extreme examples are the spermatogonia in mites which have only two divisions (Sokolow, 1934) and in fishes which may have as many as 14 (Billard, 1969; Holstein, 1969). In various animals, the successive spermatogonial generations have been named in various ways which are compared in Table 2. Presumably all spermatogonia, but certainly the secondary ones, are differentiating cells, and their differentiation is somehow connected with the process of mitoses. Successive generations in interphase show characteristics progressively different from the former generations (Fig. 2). Apparently the nuclear condition changes with each successive mitosis closer to a pre-condition for the meiotic divisions. The evidence for this is plausible but not conclusive.

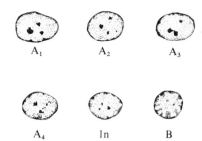

Fig. 2. Nuclei of spermatogonia of successive generations. (After C. Huckins (1971). *Anat. Rec.* **169**, 533–58).

It has been reviewed by Roosen-Runge (1969), and is also given in some detail in the chapters on insects and on vertebrates.

On first sight the mitotic phase of spermatogenesis seems to be the main determinant of the number of spermatozoa which an animal produces. But on further consideration this can hardly be the case. The number of spermatogonial mitoses is small in all animals with a few exceptions. In general, only an 8- to 32-fold increase in cell number is produced during this phase and there are no reports that the number can be increased by any known agent. The typical amount of sperm in each animal, therefore, is essentially determined by pre-spermatogenic cell multiplication. On the other hand, the progressive events of successive mitoses appear to be essential steps in premeiotic differentiation (see Chapter 9).

The mitotic stage of spermatogenesis is followed by the ''meiotic stage'' which is entirely taken up by two meiotic or maturation divisions. The cells in this stage are called *spermatocytes.* A cell during the first division and its elaborate prophase is a *primary spermatocyte* or *spermatocyte I.* The second division follows rapidly upon the first, sometimes without a proper interphase. During this period the cells are called *secondary spermatocytes* or *spermatocytes II.*

The essential event of the meiotic stage is the reduction of the diploid (somatic) set of chromosomes to the haploid complement which the spermatid and the spermatozoon carry. Because of its pivotal significance for the processes of sexual reproduction, the mechanisms of meiosis are explained in detail in every textbook of general biology, cytology or histology. Here they will be only summarized briefly (Fig. 3).

Before the first meiotic prophase begins, DNA synthesis, the doubling of the DNA strands necessary for both the following divisions, takes place. When the chromosomes first make their appearance as thin threads in the *leptonema* (or *leptotene*) stage their number is still diploid. The homologous chromosomes then become paired and closely applied to each other in synapsis in the *zygonema* (or *zygotene*) stage. The pairs twist around each other and appear like threads in the *pachenema* (or *pachytene*) stage. Because of the pairing the number of the threads corresponds to the haploid number of chromosomes. During pachenema each

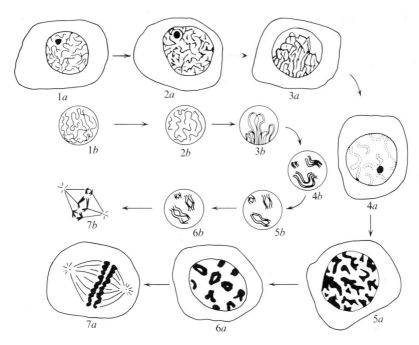

Fig. 3. Generalized drawing of successive nuclear stages of the first meiotic division. Sequence (*a*) semi-diagrammatic; sequence (*b*) schematized.

chromosome splits lengthwise to form two chromatids, but these still adhere firmly to each other. While the chromatids are in close proximity breakages occur along their length and sections from different members of chromosome pairs become joined. This is the process of *crossing over* by which hereditary material of the two homologous chromosomes is redistributed in the four chromatids adhering in synapsis. The complex unit of four chromatids representing two chromosomes is called a *bivalent*. The splits between chromosomes and often also between chromatids become visible during the subsequent *diplonema* (or *diplotene*) stage, but the bivalents still remain coherent because of the *chiasmata*, or crossings, which have been created between chromosomes through the crossing over process. As the diplonema progresses, the chromatids contract and thicken more and more, and are then ready for the actual divisions in the stage of *diakinesis*. In anaphase the two chromosomes forming a bivalent separate and move to opposite poles of the cell. In the second divisions the chromatids of each chromosome separate and as a result, each of the four cells originating from one primary spermatocyte possesses one chromosome for each pair of chromosomes present in the spermatogonia. The haploid condition is achieved.

The chromosomal changes during meiosis are intricate and variable. They have been studied for many decades and in many animals, particularly in insects and mammals. A splendid review of details is contained in E. B. Wilson's textbook on *The Cell in Development and Heredity* (1925). Recently, remarkable progress has been made in the ultrastructural analysis of certain meiotic features, e.g. of the *synaptonemal complex* which occurs in a wide range of animals as the axis of pachytene bivalents, and parallels in its behaviour the changes during synapsis and subsequent disjunction (Moses, 1969; Moses, Counce & Paulson, 1975).

In addition to the nuclear changes there are in many cases changes in the cytoplasm, the most conspicuous of which is an increase in volume. Primary spermatocytes often grow, for instance in mammals, to many times the volume of the late spermatogonia from which they arise. Spermatocytes have, therefore, often been called *auxocytes* (from the Greek *auxein*, to increase or grow), but cellular growth is apparently not a necessary condition of meiosis. In many animals, for instance in teleosts, the spermatocytes do not increase appreciably in size. Another cytoplasmic event which may have general significance although it has been reported only a few times in mammals and arthropods, is an increase in RNA during early pachytene stages and a subsequent dramatic decrease. In fact, the cytoplasm and the nuclei of secondary spermatocytes and spermatids are relatively devoid of RNA. There are other indications that the first part of first meiotic divisions is a time of vigorous cell metabolism (Roosen-Runge, 1962).

After meiosis, the last stage of spermatogenesis, *spermiogenesis*, begins. The haploid gamete during this phase of its life is called a spermatid. Immediately after the second meiotic division the spermatid appears like a relatively small, undifferentiated cell. It does not divide again, although there may be a very few exceptions to this rule (see Roosen-Runge, 1962). In contrast to spermatogonia and spermatocytes, which usually show only minor and often subtle differences between species, the spermatid usually undergoes spectacular development as species-specific features develop. Many reviews have been published on the cytology of spermiogenesis, e.g. Nath (1966), Fawcett (1958, 1970), Phillips (1974). Examples for the interesting and varied aspects of ultrastructural detail in spermiogenesis are also found in Baccetti (1971). The details of spermatid differentiation vary with the shape of the nucleus, the type of acrosome, the specific structure of the flagellum, etc. It is difficult to describe a general type of spermiogenesis in the animal kingdom. Instead, we will briefly survey the types of spermatozoa which represent the outcome of various types of spermiogenesis. The intricacies of the field of 'spermatology', i.e. the science of spermatozoa and their differentiation, are beyond the scope of this book.

The following account is largely based on Franzén's (1956) extensive survey of invertebrate spermatozoa and on Afzelius' (1972) more general, brief discussion. The so-called 'primitive type' of spermatozoa is most widely distributed in the animal kingdom. It is represented by small, flagellated cells with a spheroid or conical nucleus and few, usually four, mitochondria surrounding the subnuclear region of the flagellum. The differentiation of the flagellum is the most conspicuous feature in the development of this type. The mitochondria usually become modified in shape and larger in volume during the course of spermiogenesis. There is also commonly a sloughing of cytoplasm which serves to reduce the cell volume and the organelles. The primitive type appears to have originated early in evolution and is adapted to external fertilization in water. It is found in all animals in which this mode of fertilization prevails, in sponges as well as in many fishes and in many species of almost all invertebrate phyla. Wherever, the primitive type occurs in animals with *internal* fertilization, as for instance in some molluscs and tunicates, conditions appear to be similar to external fertilization.

'Modified types' have developed from the 'primitive type', for instance in mammals, presumably as adaptions to special conditions of fertilization. Features which are modified characteristically are: elongated nuclei ('heads'), or nuclei of highly specialized shapes, mitochondria in increased numbers and special configurations (the 'mitochondrial sheath' in mammals) and undulating membranes. Such altered features have arisen in many stages of evolution and often in isolated instances, so that one of two closely related species may have 'primitive', the other 'modified' spermatozoa. There appears to be no instance in which the adaptive advantage of a particular modification is clearly understood.

A third and relatively infrequent type is the 'simplified spermatozoon' without flagellum which occurs in a variety of bizarre forms in some crustaceans, nematodes and other invertebrates. As one might expect, in these cases spermiogenesis is also of a very special nature. In general it can be stated that the spermatozoal type, and therewith the details of differentiation of the spermatid, is not directly correlated with the evolution of the animal but with the intricate and varied details of the process of spawning or mating.

When does a spermatid become a spermatozoon, or how can the end of spermiogenesis and therewith of spermatogenesis be defined? This simple question has no simple answer. For mammals it has been generally agreed upon to call a cell a spermatozoon when it is released from the seminiferous epithelium and becomes a 'free' cell. This event is easily identifiable and the definition is, therefore, convenient, but it is well known that the newly formed spermatozoon is neither morphologically nor physiologically completely matured. The maturation process continues while the cell is on its voyage through the male accessory tract and, perhaps, until the very moment of fertilization. In other animals, germ cells are released from the gonadal tissue at stages of much greater immaturity, sometimes just after the spermatogonial divisions, and in these cases it is impossible to define the end of spermiogenesis other than by the time at which the spermatid has come to appear like a 'mature' spermatozoon, although here again the cell which appears mature may not yet be capable of fertilization.

The foregoing description of spermatogenesis has been given, for didactic reasons, as though the germ cells were living singly in isolation, whereas in reality they live in inseparable interdependence with other germ cells and usually, though not always, with certain specialized somatic cells. It is questionable whether gonocytes after the earliest embryonal stages (primordial germ cells) ever lie singly

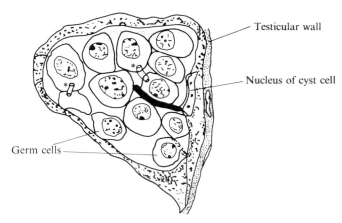

Testicular wall

Nucleus of cyst cell

Germ cells

Fig. 4. Spermatogenic cyst of the honey bee. Asterisks are at intercellular bridges. (After F. Meves (1907). *Arch. mikrosk. Anat.* **70**, 414–91.)

in an indifferent somatic tissue, but certainly after the first spermatogonial division, i.e. in the very beginning of spermatogenesis, the gametes develop in syncytial clones in almost all instances. After mitosis spermatogonia separate incompletely and remain connected by cytoplasmic "intercellular bridges". While older reports of such connections are often equivocal because of the limits of resolution of the light microscope, recent investigations with the electron microscope have validated many old data and have added much new evidence on the widespread occurrence of incomplete spermatogonial cell divisions. It has been shown that as the consequence of this mode of mitosis large interconnected clones of spermatogonia may arise, their size depending on the number of divisions which the progeny of single primary spermatogonia undergo in any given species. Cell divisions in these clones are always synchronous. There are equally as many convincing demonstrations of intercellular bridges between spermatocytes and between spermatids and in many cases clones of hundreds of spermatids appear to form "syncytia", although it is technically difficult to demonstrate this quantitatively.

There has been much speculation about the purpose of these clonal syncytia which are so characteristic for the spermatogenic process. To say that the incomplete cell divisions promote synchrony is only stating a plausible causal connection, but nothing is definitely known about the advantage to be gained by the synchrony. A solution to this problem may emerge once the role of the somatic cells in their relationship to the clones becomes more thoroughly explored. Throughout the animal kingdom there is a strong tendency for accessory cells to become intimately associated with the developing gametes, although in some invertebrates, particularly those in which a major part of spermatogenesis takes place within the coelomic fluid, no accessory cells have been found. The association may take on a variety of forms which are discussed comparatively in Chapter 11. A relatively simple example is the 'cyst' in the testis of insects (Fig. 4). Here an early

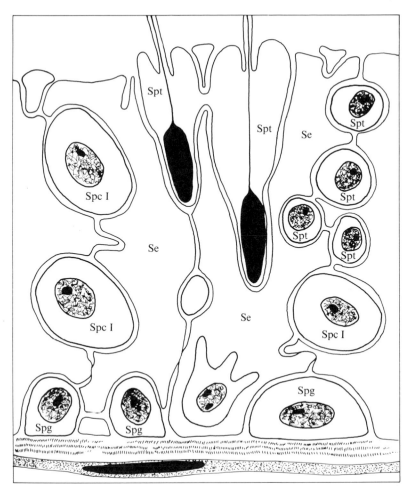

Fig. 5. Generalized drawing of mammalian Sertoli cells and germ cells. Se, Sertoli cell; Spg, spermatogonia; Spc I, spermatocyte I; Spt, spermatid. (After L. Nicander (1967). *Z. Zellforsch. mikrosk. Anat.* **83**, 375–95 and M. Dym & D. W. Fawcett (1970). *Biol. Reprod.* **3**, 308–26.)

spermatogonium becomes surrounded by one or more 'cyst cells' which remain the envelope for a single clone throughout the stages of spermatogenesis and often show changes indicating that their metabolism is correlated with that of the gametes. Another well known and more complex example is the 'Sertoli cell' (Fig. 5) of mammals (Fawcett, 1975). These cells are dispersed in a regular pattern among the germ cells of the seminiferous tubules and extend from the basement membrane through the whole thickness of the germinal tissue. Throughout a continual cycle of developing germ cells, each Sertoli cell is in close association with a definite number of germ cells in various stages. The differentiating germ cells move along the surface of the Sertoli cells, as on an assembly line, from the periphery toward the lumen of the tubule. While the spermatogonia are not completely surrounded by the peripheral pillars of the Sertoli cells, spermatocytes and spermatids are enveloped by Sertoli cell processes and the Sertoli cell surface shows modifications in fine structure where it borders the heads of spermatids in later stages of differentiation. All data point to a symbiotic relationship between gametes and 'supporting' or 'nurse' cells. This is not a simple association. No doubt it serves to condition the supply of nutritives for the spermatid (see Peter, 1899; Chapter 2). It is also certain that the Sertoli cells phagocytize the 'residual bodies', i.e. the sloughed cytoplasmic parts of spermatids. More than that, the obvious multitude of interrelationships between the nurse cells and spermatogonia, spermatocytes and spermatids makes it appear probable that the close correlations between stages of germ cells in the cycle of the seminiferous epithelium are mediated through the nurse cells.

In the preceding general and basic survey I have presented an outline and some necessary terminology of spermatogenesis for the student wishing to be initiated. But more importantly, I have attempted to establish viewpoints – my own – from which the compilation of data in the remainder of the book becomes understandable and perhaps interesting. I see spermatogenesis as a basic biological process similar in some ways to early embryogenesis, a process of development not simply of a multitude of individually proliferating and differentiating cells, but an organismic event throughout which the developing cell populations are interdependent and integrated. Today, we do not know specifically what constitutes the essential uniqueness of a male gamete, we do not know what enables it to perform reduction divisions and what common characteristics except genes predispose it to differentiate into a special 'gene-carrier' capable of joining its female counterpart. This book, of course, does not provide the answers. It does, however, enable the reader to gain an insight into the variety of cellular communities which appear to produce specific spermatozoa in a closely integrated process. As long as the relevance of single data to the main problem remains uncertain, a survey must present a great deal of potentially superfluous, although possibly helpful, material, but among the multitude of items in its dragnet there must be some which will turn out to be centrally significant. By attempting to cover the whole animal kingdom as well as possible, I hope also that investigators will be led to species which will most readily answer their basic questions.

2

Historical aspects

The approaches and the results of research are in the broadest sense inseparable from history. A great part of this book consists of reviews and is by its very nature oriented historically. In this chapter an attempt is made to gain a perspective in which the many individual references may be viewed. The chapter is the result of my fascination with the minds and concepts which dominated certain periods of biological research during the last 140 years. As such it is a side issue and not a necessity. Its main purpose is to stimulate the reader to contemplate reasons for the peculiar patchwork of scientific data and projects with which an investigator in the field today is confronted. Spermatogenesis, like other basic biological problems, has been investigated in a rambling fashion depending on the personalities of the scientists involved, on the methods accessible to them, and particularly on the animals they happened to use as objects for study. At any given time in history, the sum total of activity in the field has produced a climate of opinion which characterized and determined the individual investigations and the concepts which developed from them. Our present epoch is no different in this respect, and an exploration of the character of circumscribed periods of the past may serve to illuminate the significance of the present condition.

The 140 years of history to be surveyed may be divided into three periods according to the main scientific direction, emphasis and methods employed. The first was the immediate result of a revolutionary new insight, the cell theory, which promised to open a new understanding of biological structure. With regard to spermatogenesis the validity of the cell theory was fully established in this period, the first 25 years after the publication of Schleiden's and Schwann's work. During this time almost all results were achieved by microscopy on fresh tissues with a minimum of technical manipulation. In the 1860s began the second, long period of painstaking but, in view of the whole animal kingdom, very unsystematic exploration of the morphological details of spermatogenesis. It lasted about 90 years. Progress was achieved mainly through the application of histological and cytological techniques with some assistance from cytochemical methods and tissue culture. Large amounts of data were accumulated and concepts became more detailed and sophisticated but hardly any new concepts originated. The third period, our own, began in the 1950s with a new emphasis on the kinetics of cell populations. Immediately thereafter the electron microscope came into general use. The result was a further rapid increase of more and more detail in smaller and smaller dimensions, together with a shift of emphasis to larger views in which germ cells are seen as interdependent with their micro- and macroenvironment and at the same time an understanding is sought of the molecular processes determining cellular behavior.

The beginning of research on the development of male germ cells coincides almost exactly with the advent of the cell theory which naturally led to the discovery of the "cellular" nature of 'spermatozoa' (the word appears first in Baer, 1827). However, it is not quite clear who first enunciated the idea that spermatozoa are derived from cells of the testis. Koelliker (1841) who collected the first large body of evidence on this issue, maintained that it was Rudolph Wagner (1836) whose illustrations (1839) are shown in Plate 1. Wagner examined fresh fluid from mammalian testicular tubules under the microscope and found in them 'peculiar granules or spherules of a very variable shape and size. The proportion of these granules to the amount of fluid is approximately that of blood corpuscles to the blood, or pus corpuscles to the pus.' He also saw 'Samenthierchen', sperm animalcules, spermatozoa. Wagner felt certain that some of the various types of spherules were forerunners or preceding stages of spermatozoa. His work, in which the word "cell" does not occur, is characteristic of the type of concept formation which immediately preceded the cell theory.

A few years later Koelliker made his fundamental observations. Their circumstances are exceptionally well attested because Koelliker, then a medical student, 23 years old, described them in letters to his mother while the research was in progress. He published them in his memoirs in 1899. The episode is so characteristic and throws such an intimate light on the makings of scientific discovery that it will be retold here in some detail.

In 1840, Koelliker whose life was to extend to the twentieth century, was at the very beginning of an illustrious career and under the influence of the excitement which the ideas of Schleiden and Schwann had stirred in the minds of his teachers, Johannes Müller and Jakob Henle, in Berlin. Every biologist was looking for confirmation of the cell theory in all conceivable objects, and Müller in particular, whose interest in animal tissues was very broad, appears to have suggested marine invertebrates as suitable objects for investigations into the cellular nature of animals. As a result, four of his students set out in the late summer of 1840 with the plan to spend a "vacation" on the North Sea islands of Föhr and Heligoland where they intended to combine swimming, fishing and hunting of shore birds with a scientific exploration of marine life. Albert Koelliker and Carl Naegeli, later a famous botanist, had borrowed a microscope, at which they took turns on alternate days. It had been built by Chevalier in Paris, and Koelliker reported to his mother that it was 'still quite all right and useful, but not really very good'.

With this instrument Koelliker directed himself initially toward a broad exploration of fresh invertebrate tissues from which he hoped to develop a topic for his medical dissertation. Every morning the students arose at 6 or 6.30 and collected material in the tide pools on the rocky shore. With the interruption of breakfast they worked until the main meal at 3 p.m. The afternoon was spent in recreational activities, but later in the evening there was more work recording and identifying specimens, discussing them and reading. When after 7 weeks on Föhr the "vacationers" moved to Heligoland for another 16 days, Koelliker had a tentative title for a doctoral dissertation: 'Beiträge zur Kenntnis der Samenthiere niederer Thiere'. The title reflects what was already Koelliker's opinion, that every male animal produces spermatozoa characteristic of its species, but on his return to Berlin he seems to have discovered that he needed to establish also that male sex occurs in all invertebrates. He proceeded to spend the greater part of the nights of the following semester, the one before his final examinations for the degree of a medical doctor, in searching through libraries and restudying his material with a better microscope. The result of this period of concentrated work was a treatise of 88 pages finished 5 months after Koelliker had examined the first specimens for his investigation: 'Beiträge zur Kenntnis der Geschlechtsverhältnisse und der Samenflüssigkeit wirbelloser Thiere, nebst einem Versuche über das Wesen und die Bedeutung der sogenannten Samenthiere.'

This paper established an astonishing number of the most fundamental facts pertaining to spermatogenesis. It is worth remarking that this foundation, which has proved its validity until today, was laid in a principally sound and systematic manner through a broad survey of invertebrate animals. Thirty-seven species were investigated including coelenterates, molluscs, echinoderms, annelids and many crustaceans. The details of the observations make interesting reading even today, but only the most salient points will be mentioned here. In the second part of the treatise Koelliker reviewed 'present day knowledge concerning the semen of animals', and made the following points:

(1) In the semen of all animals, with a few exceptions, motile particles ('Theile') are found, the 'Samenfäden' (spermatozoa).

(2) The spermatozoa are the most essential part of the semen.

(3) The spermatozoa develop singly or in bundles *from* or *in* cells which form at times of sexual maturity or activity in the testes through processes analogous to the development of cells ('thierische Elementartheile') but significantly different from the development of animals out of zygotes. For many years Koelliker tended to think that spermatozoa develop from nuclei rather than from whole cells but in his first treatise he did not mention this point.

(4) The shapes of spermatozoa are limited in variety. Usually they are similar within genera, frequently also within families and classes. Each animal appears to have only one kind of spermatozoa, but Koelliker referred to the snail *Paludina* as a notable exception.

(5) Spermatozoa do not have the organization of Metazoa and do not reproduce their kind.

In his conclusion, Koelliker advocated that the view of the spermatozoa as the earliest condition of a developing animal, held by Leeuwenhoek and many others, should be abandoned. The spermatozoa should not be considered accidental inhabitants of the semen, animals of a separate kind (parasites), as Cuvier, Baer and others had believed. Koelliker mentioned that Müller, Henle, Dujardin and Prevost shared his view of spermatozoa as essential cellular parts of the semen, but there is no doubt that it was he who brought forward in his dissertation the bulk of the evidence which gave impetus to the detailed histological discoveries later in the century. Koelliker remained active in research on spermatogenesis for another 15 years. His last paper on the subject (1856) dealt with physiological experiments on spermatozoa. In it he expressed the belief that the head of the spermatozoon is the part essential for fertilization, but he still did not recognize a contribution from the cytoplasm to the developing sperm cell. In the face of an almost total lack of biochemical facts he went as far as he could in experimenting with the reactions of spermatozoa to a whole host of chemical and physical conditions. One has the impression that his disappointment over the meager results made him abandon this field of interest for the rest of his long life.

The years of Koelliker's active participation approximately circumscribe the first period of research on spermatogenesis during which the methodology was limited to microscopic observations on living or fixed and unstained tissues. During this time little work was done on tissues such as mammalian seminiferous tubules which could not be made to survive easily. A good example of an object which could be clarified by observations on living tissues alone is the work of Lieberkühn on spermatogenesis in freshwater sponges (1856), which is discussed in some detail in Chapter 3 (see also Plate 4a). At the borderline of a new era brought about by histological and cytological methods stands the work of Sertoli (1865) who demonstrated the supporting cells of the human testis with remarkable success, not by sectioning and staining, but by careful and slow maceration techniques.

From that time onwards the exploration of testicular tissues made rapid progress as a host of scientists experimented widely with fixing, sectioning and staining techniques. The work of Schulze (1875 and later) on sponges, of Valette St George (1865 until 1887) on vertebrates, Sars (1867, as quoted by Retzius, 1909) on crustaceans, Salensky (1907) on flatworms, Ebner (1871 and later) on the rat, and many others, exemplify the early decades of the histological era which reached its full development in the years from 1885 to 1910.

In 1865, Valette St George began publishing a long series of investigations on fresh and fixed spermatogenic tissues of many vertebrates, which resulted in the gradual identification of different cell types as stages of spermatogenesis. The terms spermatogonium, spermatocyte and spermatid were created by Valette St George, but his terminology was not very clearly defined and understood for several decades. A contemporary discussion of the confusion which prevailed for a long time may be found in Ebner (1888), and a modern attempt at clarification in Roosen-Runge (1962) and in the Glossary of this book. The most important discovery within the spermatogenic tissues was the demonstration of the supporting cells in human seminiferous tubules by Sertoli (1865). The recognition that special auxiliary cells may contribute to the development of gametes stimulated interesting and significant discussions and reviews. Because these foreshadowed many of the concerns of this book, they will be described here in some detail.

Waldeyer (1887) in a talk to the first assembly of the Anatomische Gesellschaft at Leipzig (under the chairmanship of Koelliker) reviewed for the first time the nineteenth-century phase of research on spermatogenesis. Although his paper was written at the time when cytological phenomena of gametogenesis were beginning to attract great interest, Waldeyer did not emphasize these aspects of sperm development but concentrated on the relationship of germ cells with each other and with the supporting cells in the testis. The central event in spermatogenesis, meiosis, Waldeyer did not even mention, although the existence of heterotypic divisions had been clearly postulated in the early 1880s in Beneden's work on the behaviour of the nuclei in the fertilization of *Ascaris* eggs. The term "meiosis" did not come into use until the early twentieth century. It was first introduced, spelled 'maiosis', in a paper containing excellent drawings of spermatocytes of the cockroach (Farmer & Moore, 1905).

In 1887 it was still acceptable to classify spermatozoa in animals simply as cell-like ("zellige") or thread-like ("fädige"). Cellular details did not go beyond fairly gross parts such as "head", "tail", "middle-piece" and "axial thread". For the stages of germ cell development Waldeyer proposed the acceptance of Vallette St George's terminology which has been in use ever since. In the second part of his talk he reviewed the evidence pertaining to the presence of *two* types of cells in the male germinal tissue of vertebrates and of some invertebrates. The confusion in the interpretation of available data had just reached a peak. While Sertoli had demonstrated in exemplary fashion that in the human testis 'round' germ cells coexisted with 'branched' supporting cells, Ebner (1871) had suggested that the round cells were leucocytes and the dendritic cells the progenitors of germ cells. Biondi (1885) had surmised that the branched cells were formed by the cytoplasmic remains of spermatids, and Benda (1887a) in a talk at the same meeting at which Waldeyer spoke, stated that the round and the branched cells 'copulate' with each other and thus lead to the formation of spermatozoa. To us such confusion appears almost ridiculous but it reflects the fact that the morphological nature of the symbiosis between germ and supporting cells cannot be resolved with the light microscope.

Waldeyer realized how incompletely the animal kingdom had been explored, and strongly suggested for the future a systematic and comprehensive comparative investigation of invertebrates and also of lower vertebrates. This plan, of course, has never been realized, and unsystematic and fragmentary investigation of the spermatogenic process has continued. However, the general concepts were more thoroughly debated toward the end of the nineteenth century than at any other time, and several attempts at synthesis were made before the century closed.

Prenant (1892) based a review on Waldeyer's paper, under the title 'Sur la signification de la cellule accessoire du testicule et sur la comparison morphologique des éléments du testicule et l'ovaire.' He surveyed all phyla on which any relevant work had been done and quoted more than 100 references. He cited evidence for the existence of accessory cells in vertebrates, tunicates, arthropods, molluscs, and sponges, without considering the question of whether the observations justified regarding all "supporting" tissues as homologous. With flourish he espoused the hypothesis that the egg and the male supporting cells on the one hand are homologous to the spermatozoon and the follicular cells on the other. Therefore, he called the supporting cells in the male 'female elements of the testis' and the follicular cells 'male elements of the ovary'. This fanciful terminology caused considerable confusion in the contemporary literature. It was reinforced through the writings of Gilson (1884 etc.) who named the nuclei of testicular supporting cells 'female nuclei' even though he was not wholeheartedly an adherent of Prenant's hypothesis (see Chapter 6.1).

In 1899, Peter published a brilliant review and some interesting speculations on the process of spermatogenesis. He produced evidence that nurse cells existed in the testes of teleosts where they had not been demonstrated previously. He divided invertebrate testes into three groups, those with and without nursing elements, and those with anuclear nutritional material. Prenant had stated that arachnids, platyhelminthes, echinoderms and medusae did not possess nurse cells. Peter removed *Ascaris* and some echinoderms and medusae from that list and suggested that on closer inspection many more species might be shown to have nurse cells. On the significance of the nurse cells he quoted Gilson's (1887) idea that sperm cells transform so rapidly that they cannot supply their own needs and must have auxiliary cells to nourish them. Peter himself went further. He believed that the mode of transformation of the spermatid is the specific reason why these cells assimilate nutrient 'at first only with difficulty, in later stages not at all'. He came to the enlightened conclusion that the particular state of the chromatin is responsible. 'We have every reason to believe that chromatin in its compact form, which it assumes during spermiogenesis, suitable for travel so to speak, is not from the point of view of nourishing the cells, and vice versa. In brief, the condition of the nucleus which we usually consider active is one of complete rest from the point of view of nourishing the cells, and vice versa, In brief, the condensed form of chromatin with a small free surface is incompatible with the processes of nutrition and synthesis.' For Peter, the nurse cells were a *conditio sine qua non* in the transformation of germ cells and it is not surprising that he believed that the process of spermatogenesis must be essentially similar throughout the animal kingdom.

These reviews expressing what scientists regarded as the gist of the data collected during the century, did not reflect the great change which was even then in progress through the coming into flower of cytology. The pursuit of fine cellular detail, often near the limit of microscopic resolution narrowed the view of many investigators to very special phenomena. For decades it appears to have been impossible to integrate the multitude of specific data into the general picture of spermatogenesis. Cytology and cytogenetics influenced research in the field during the first half of the twentieth century to such an extent that the focus shifted to the chromosomal details of meiotic cell divisions and of cytological differentiation of spermatids. This is beautifully demonstrated by E. B. Wilson's authoritative book "*The Cell in Development and Heredity*" (1925) in which the most exhaustive review on research in spermatogenesis is presented in the chapters on gametes and meiosis. There have been no other comprehensive reviews on spermatogenesis in animals until very recently (Roosen-Runge, 1969, 1974). The rapidly accumulating results of the exploration of cellular detail in male gametes greatly advanced the knowledge of chromosomal structure and behavior but did little to expand an understanding of germ cell populations. However, throughout the decades a relatively slow but steady stream of contributions appeared describing the general features of spermatogenesis in additional species. Outstanding examples are the following:* Regaud (1901; 8) on the rat, Tönniges (1902; 6.1) on a centipede, Ancel (1903; 5) on the vineyard snail, Depdolla (1906; 7) on an earthworm, Demandt (1912; 6.2) on the water-beetle, Kirillow (1912; 8) on the horse, Fauré-Fremiet (1913; 7) on *Ascaris*, Lindner (1914; 7) on a parasitic flatworm, Buder (1917; 6.2) on a butterfly, Turner (1919; 8) on the perch, Dalcq (1921; 8) on a slow-worm, Branca (1924; 8) on man, Ankel (1924; 5) on a marine snail, Merton (1930; 5) on the ram's horn snail, Sokolow (1934; 6.1) on spiders, Newby (1940; 7) on an echiuroid worm, Coe (1943; 5) on the ship worm and other molluscs, Anderson (1950; 6.2) on the Japanese beetle, and Matthews (1950; 8) on the basking shark. These arbitrary but very representative examples indicate something of the slow pace with which the animal kingdom has been explored. Despite the great variety of findings, general biologists throughout this period adhered to the opinion that spermatogenesis is very much alike in all animals. A characteristic example is L. Hyman's (1940, p. 431) statement with regard to *Hydrozoa*, that young spermatogonia develop into spermatozoa 'by a typical spermatogenesis' (see Chapter 4).

The third and last period of spermatogenic research is our own. It is contemporary history and I cannot undertake to give an "objective" account of it because I myself was a small actor on the crowded stage. The period began for me with a conference under the auspices of the New York Academy of Sciences (see Leblond & Clermont, 1952; Roosen-Runge, 1952) on the biology of the testis. At this meeting, which concerned itself exclusively with the male gonad of rat and human , it became apparent first that the emphasis in spermatogenic investigations was shifting and that new methods were becoming available. It was first of all the dynamics of germ cell development which was discussed, and as the subject was first explored in mammals which are well suited for this approach, the kinetics of cell populations of the testis has, in the last 20 years, become much better known in mammals than in any other group of animals. However, results in insects and in teleosts have shown that quantitative comparison can reveal significant insights

---

* The numbers which follow the year in these references indicate in which chapter's reference list the references are to be found.

into the general and specific features of the spermatogenic process. The advent of the technique of labeling with [³H]thymidine those nuclei which are in the process of synthesizing DNA has greatly furthered the study of cell kinetics and has produced data on the duration of cellular processes.

In addition, the electron microscope has in the last 15 years extended resolving power far beyond that of the light microscope so that intercellular relations at cell borders are revealed and cellular organelles, for instance during the complex transformations of the spermatid, can be studied in a new dimension of detail. In turn, the high resolution of morphological features serves to define and correlate stages of cellular kinetics.

The modern approaches, greatly accelerated by the increase in professional scientific manpower since the Second World War, are producing a tremendous wealth of new data which are hard to digest and correlate by themselves, but which in addition necessitate a new look at the data and concepts of preceding periods. Many results of a former age need to be reinvestigated. A large number of the papers referred to in this book contain fascinating suggestions to be explored with new methods and new views. The lesson to be learned from this casual historical review is probably not that a much more systematic coverage of the animal kingdom is needed. A general systematic approach is not practical in our culture, even if it were desirable, and the increased amount of scientific activity will improve coverage even when the pursuit is individualistic and unsystematic. But it seems clear that the coverage of the last 140 years does not entitle us to make strong and confident generalizations about the process of spermatogenesis. Basic biological processes are adaptive and expressed very variously in different animals. In some they are far more open to scientific manipulation and inspection than in others. With the deficiencies and omissions of past exploration firmly in mind, the incentive should be great to remedy them.

We have seen that all knowledge of spermatogenesis has been acquired as a logical consequence of an initial conceptual stimulus, the cell theory, which called for the *cellular* analysis of living beings. No similarly forceful stimulus to progress in biological research has occurred in the last 135 years. But there are large outside forces discernible today which may serve to strengthen the vigor which is devoted to the study of male reproduction. The steady and rapid growth of awareness of the problems which the world community is facing, the recognition of the threat of overpopulation and of the need for birth control, have led to energetic and practical approaches to fertility and sterility in man (Fawcett, 1975). In medicine it has been in the past the *female* reproductive system which for many decades received a very large share of attention. The field of "andrology" has become a small counterpart of gynecology only in the last 10 years, but this slight change in emphasis already has had a significant impact on research of mammalian male gonads and spermatogenesis. Examples are the volumes which recently have begun to proliferate on the subject (e.g. Johnson, Gomes & Vandermark, 1970; Rosenberg & Paulsen, 1970; also the journal *Andrologia*, 1974, formerly *Andro-*

logie). The new direction is overwhelmingly oriented toward human problems in the broadest sense, and is led by investigators in the medical and paramedical fields. Certainly it will produce significant additions to our knowledge of human spermatogenesis. In the presence of this thrust, it is all the more necessary to keep the fundamental biological significance of the subject firmly in mind. Our consciousness of the world is increasing not only with respect to mankind, but with regard to all life on earth, and features of human biology must be seen as expressions of underlying principles which become understandable through intensive and broad comparisons among the host of living beings.

3

Porifera

The sponges, the lowest phylum of the Metazoa, are considered to be of 'the cellular grade of construction', without organs, mouth or nervous tissue (Hyman, 1940). In reviewing this phylum one may expect to find a spermatogenic process integrated with the soma in the simplest possible fashion. Here, if anywhere, the germ cells are developing in a surrounding which because of its relative simplicity may not exercise a complex control over their differentiation. Perhaps, in this phylum, the bare essentials of the spermatogenic process, the minimum conditions under which it can be successful in a multicellular organism, will be revealed. Although the male organs have been explored in less than 20 species (less than ½% of the total), a considerable variety of phenomena have come to light. However, in contrast to mammals where recent results have all but eliminated the need to evaluate the older literature, in the sponges information is still so scanty that we must utilize even the limited insights gained many decades ago. In recent years Tuzet (1964) has presented a very brief but poignant review. In Fell's (1974) review of reproduction in Porifera, 3 ½ pages are devoted to an updated survey of spermatogenesis.

In some sponges it is possible to observe the gonads in the living, although this possibility has not yet been exploited fully. Lieberkühn (1856a) looked in *Spongilla fluviatalis* for the places of origin of the 'zoosperm-like corpuscles' (i.e. spermatozoa) and found them in the form of 'spherical containers enveloped by a structureless, transparent membrane, which are surrounded by cells of the sponge. Their diameter is approximately 1/12 mm'. In an addendum (1856b) he stated that he had seen some of the containers or capsules filled only partly with motile 'Spermatozoidea', and, in part, with structures which gave rise to sperm (Plate 4a). In the 1870s the sponges were a favorite subject of zoologists and anatomists, Haeckel (1871) found spermatozoa singly in the entoderm, their tails sticking out of the epithelium, but Eimer (1872) contradicted him sharply and described balls of 'millions' of sperm cells, scattered through the tissues. Both authors, however, agreed that the cells from which sperm originated were the choanocytes, i.e. differentiated entodermal cells. At the same time there were many investigators who still did not believe that sponges had any spermatozoa at all, a view which only gradually faded under the impression of Schulze's series of studies on the structure and development of sponges (1877, 1878, 1879, 1881). In several investigations Schulze described spermatogenesis in a number of genera, almost all belonging to the horny sponges. In *Halisarca* and *Aplysilla* he found clumps of

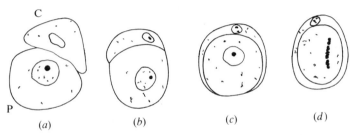

Fig. 6. Semi-diagrammatic representation of a ''primordial germ cell'' or ''stem cell'' (P) being surrounded by a ''covering cell'' or ''cyst cell'' (C) in *Spongilla*. In (*d*) a cyst has been established and the first spermatogonial division is taking place. (After W. Görich (1904). *Z. wiss. Zool.* **76**, 522–43.)

male germ cells, of greatly varying sizes. He concluded that each clump originated from a single cell, an amebocyte, by continuous division. The capsule which surrounds the clumps after the very early stages of development he described as consisting of polygonal, flat, 'endothelium-like' cells. In general, Schulze tended to the view that each clump contains germ cells in the same stage of spermatogenesis. He stated this expressly for *Hircinia spinulosa* (1879).

Polejaeff (1882) discussed in detail two of the issues which have since that time emerged again and again and not only with respect to sponges, namely, the origin of male germ cells and the origin of the cells delimiting the germinal epithelium from the rest of the body. He described that in *Sycandra raphanus* each amebocyte divides into a primordial germ cell ('Ursamenzelle') and a covering cell ('Deckzelle') which immediately assumes a peripheral position (Fig. 6). The covering cell, according to Polejaeff, does not divide again, but continues to enclose the division products of the germ cell. He recognized that most siliceous and horny sponges described by Schulze and others had a different mode of follicle formation in which a covering of flat capsule cells developed secondarily from the surrounding mesenchyme, and he speculated on the mechanism involved. Rejecting the simple explanation that growth pressure from the developing sperm masses might compress and arrange the surrounding cells, he suggested that the mesenchymal cells had acquired an inherent 'hereditary' tendency to flatten and surround the spermatogenic cells because such a response is obviously useful 'for the good health of the sexual product'. He further pointed out that by contrast the voluminous eggs of *Sycandra* do *not* become surrounded by a cellular envelope but remain motile until late in development. On the other hand, the small non-motile egg cells of *Spongilla* are enveloped by epitheloid cells. According to Polejaeff and most later investigators, the capsule cells show no signs of growth or differentiation during spermatogenesis, and their fate is unknown. Recently, Tuzet & Pavans de Ceccatty (1958) have reported that in *Hippospongia communis* the squamous 'follicular' cells may harbor an iron-containing pigment, probably melanin. This is so far the only direct evidence of a specific metabolic function of the capsule.

Polejaeff also opened the discussion of the question of whether the early cell divisions of germ cells are complete or incomplete. He insisted that they are incomplete in *Sycandra* and that binucleate cells result. Today, when electron microscopy has revealed clearly that dividing spermatogonia, as well as spermatocytes, often divide incompletely and remain interconnected by more or less extensive intercellular bridges, these findings of early microscopists do not appear particularly dubious. It is certainly true that all descriptions of light microscopic observations which involve the behaviour of cell membranes are subject to grave doubts because of insufficient resolution. This holds particularly for observations on fixed and stained sections. On the other hand, the best microscopic evidence obtained from whole cells in living or fresh tissues has now been confirmed repeatedly by electron microscopy. Polejaeff's observations were probably sound, as were Valette St George's (1865) who had seen and depicted interconnected mammalian germ cells in fresh tissue.

For many decades after Polejaeff's publication the story of the male gonad in sponges was occasionally enriched by new details in various species, but there was no searching, comparative discussion of the findings. Through Fiedler (1888) and Görich (1904) it became known that more than one cell may enter into the formation of the follicular envelope, for instance in *Spongilla* where up to four cells may surround a group of germ cells. Fincher (1940) found eight nuclei in the covering of the 'sperm ball' in the demosponge *Stylotella heliophilia*, and his description of what he called 'cyst formation' differed from previous ones. It appears that in *Stylotella* clumps of archeocytes in the mesenchyme become secondarily surrounded or subdivided by mesenchymal cells which form an envelope. Only after a group of archeocytes has in this way become isolated from the environment, does it proceed in spermatogenic development. The 'sperm ball' does not arise from a single cell and the development of the spermatogonia within it is not synchronous.

Another unique story was told by Dendy (1914) who became convinced that in the calcareous sponge *Grantia compressa* the choanocytes directly transform into ameboid germ cells which later divide and form 'morulae' of 16 cells, apparently not encapsulated. These are liberated and transferred by currents into the cavities of another sponge. But according to Gatenby (1919), Dendy later withdrew many of his statements at a meeting of the Linnean Society of London. Gatenby described in *Grantia compressa* nests of tailed spermatids surrounded by a capsule of mesogloeal cells which in his illustrations appear to be cuboidal and lie in the interstices between the gastral chambers. He was extremely skeptical about the work of previous investigators and maintained that 'Haeckel, Polejaeff and Dendy have failed to bring convincing evidence as to the spermatogenesis [in sponges], and no cytologist acquainted with these questions would identify the "sperm balls" of these workers as authentic stages in the formation of spermatozoa in any animal'. He completely disposed of Dendy's notion of 'sperm-morulae'. On the other hand, he considered Görich's observations valid and declared that 'Haeckel was correct in tracing the origination of germ cells from collar cells'.

Investigation of calcareous sponges is relatively difficult and the controversy raised by Gatenby has not been decided by reinvestigation in recent time. Fell (1974) simply states that the formation of cysts 'has not been extensively studied in calcareous sponges'.

In one of the rare studies on siliceous sponges, Okada (1928) confirmed many of Schulze's findings, in particular the origin of spermatogenic cells from archeocytes in the mesogloea, but he found no covering cells. The freshwater sponges were reinvestigated after some decades by Leveaux (1942). He came to the conclusion that in *Spongilla lacustris* and *Ephydatia fluviatalis* spermatogonia are derived not from undifferentiated amebocytes as Fiedler and Görich had believed, but from typical choanocytes. He observed the formation of a 'follicular membrane' by a flattening of adjacent mesenchymal cells. Meiotic divisions began when the sperm cell aggregate had attained a certain volume which varied in wide limits. Sperm was liberated by simple rupture of the follicle.

The genus *Halisarca*, Schulze's main object 80 years previously, was reinvestigated by Lévi (1956). While he considered amebocytes as the cells of origin of the spermatogonia, as Schulze had done, he was no longer quite so certain: 'It is not possible to follow entire baskets in the course of transformation, and it is difficult to prove that individual choanocytes emigrate and eventually transform into amebocytes or into spermatogonia'. In other words, he found no clear evidence that spermatogonia do not develop directly from amebocytes. Lévi observed that spermatogenesis is 'absolutely' synchronous within each follicle in *H. dujardini* but not in *H. metchnikovi* where all stages of spermatogenesis may be represented in a single follicle (Fig. 7). However, the process of formation of the follicles appeared to be identical in the two species.

Lévi considered four possibilities of follicle formation, all of which may actually occur in various species:

(1) One cell gives rise to both, the cells of the wall or capsule and the spermatogonia (only example described by Tuzet (1930), *Reniera simulans*.)

(2) Follicular cell and spermatogonia are of different origin, but each line is derived from a single cell (*Spongilla* according to Görich).

(3) Follicle and germ cells are of different origin and each derived from several cells (the most frequent type among sponges; example *Halisarca*).

(4) A variant of (3) in which follicle cell and spermatogonia are of different origin, the follicle cell being unicellular, the germ cell aggregate pluricellular in derivation (*Verongia* according to Fincher (1940)).

It is characteristic of the state of exploration of the Porifera that several genera have been described to have more than one mode of follicle formation, for instance, *Spongilla* in which according to Fiedler (1888) type (1), occurs, according to Görich (1904) type (2), and according to Leveaux (1942) type (3).

In her review (1964) Tuzet reports original observations on *Hippospongia communis* in which the origin of spermatogonia may be determined 'precisely'. Choanocytes transform into spermatogonia (Fig. 8). Subsequent development of

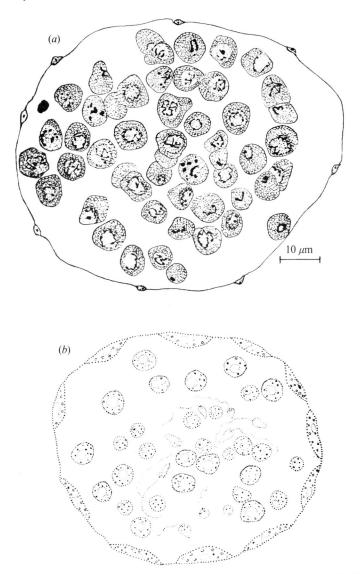

Fig. 7. Spermatocysts in the genus *Halisarca*. (*a*) Cyst of spermatocytes in *H. dujardini* Johnston. Germ cells are synchronous, but appear disconnected. (*b*) Cyst of spermatocytes in *H. metchnikovi* Lévi, drawn from life. Different stages of germ cells contained in one cyst. (After C. Lévi (1956). *Arch. Zool. exp. gén.* **93**, 1–184.)

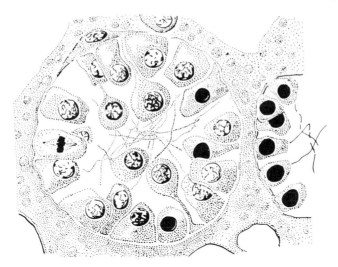

Fig. 8. A ciliary chamber of *Hippospongia communis* changing into a spermatocyst by transformation of choanocytes into spermatogonia. The chamber at right contains typical choanocytes. (After O. Tuzet (1964). In *L'origine de la lignée germinale*, ed. Et. Wolff. Paris: Hermann.)

spermatogonia in the transformed choanocyte baskets is nearly synchronous. Tuzet pointed out that there is no need to regard the origin of germ cells from amebocytes or from choanocytes as fundamentally different. Amebocytes may arise from flagellated cells of the larva, but there is also the possibility that choanocytes migrate into the mesenchyme and 'dedifferentiate' there into amebocytes which in turn may give rise to spermatogonia (Gatenby, 1919). Germ cells become identifiable at various stages of development and may originate from entodermal cells in various stages of differentiation. These facts make it impossible to accept the "continuity of germ plasm" in the classical sense as valid for the phylum Porifera.

It does not appear worthwhile to repeat the few definitive findings concerning the cytology of the developing sperm cells, which have been summarized by Fell (1974).

In conclusion, the male gonads in sponges are represented by aggregations of spermatogenic cells, which are usually, perhaps always, surrounded by a capsule of epitheloid cells. The aggregates vary greatly in size. Few measurements are given in the literature, but the order of magnitude appears to be between 70 and a few hundred micrometres. The spermatogenic foci have been called sperm clumps, or sperm balls, sperm cysts, follicles, testicles, etc. Gatenby (1919) argued that the term "testis" would be inappropriate because of the wide scattering of the sperm masses and Fell (1974) stated, I believe rightly, that sponges do not possess "true gonads". The term "testicle" as used, for instance by Tuzet, would

TABLE 3. *A list of genera of Porifera which have been investigated in some detail with regard to the male reproductive system*

Genus	Reference	Cell of origin of spermatogonium
	Class Hexactinellida; glass sponges	
?	Eimer (1872)	Choanocytes
Farrea	Okada (1928)	Choanocytes
Stylotella	Fincher (1940)	Amebocytes
	Class Calcarea; calcareous sponges	
?	Haeckel (1871)	Choanocytes
Sycandra	Schulze (1875)	Amebocytes
Sycandra	Polejaeff (1882)	Amebocytes
Sycandra	Görich (1904)	Amebocytes
Leucoselenia	Polejaeff (1882)	Amebocytes
Grantia	Dendy (1914)	Choanocytes
Grantia	Gatenby (1919)	Choanocytes
Sycon	Schulze (1878)	Amebocytes
	Class Demospongiae	
Halisarca	Schulze (1877)	Amebocytes
Halisarca	Lévi (1956)	Amebocytes
Aplysilla	Schulze (1878)	Amebocytes
Hircinia	Schulze (1879)	Amebocytes
Corticium	Schulze (1881)	Amebocytes
Reniera	Tuzet (1930)	Amebocytes
Hippospongia	Tuzet (1964)	Choanocytes
Octavella	Tuzet & Paris (1965)	Choanocytes
Spongilla	Görich (1904)	Amebocytes
Spongilla	Leveaux (1942)	Choanocytes
Ephydatia	Leveaux (1942)	Choanocytes

seem to be equally inappropriate. The characteristic encapsulation by flattened cells provides a little sac, as it were, for the germ cells and completely justifies the use of the term "follicle" or "spermatocyst" (see Glossary). The cells of this envelope, whatever their function may be, show little differentiation, and by their morphology give little indication that they participate in or react to the metabolic changes which are necessarily occurring throughout the development of the germ cells.

Nurse cells have never been found within the follicles nor has any ordered arrangement of spermatogenic cells, synchronous or asynchronous been described. Occasional illustrations (Lieberkühn, 1856a) show definite orientation of spermatozoa in relation to the wall of the follicle (Plate 4a). A similar orientation has been found by Tuzet & Pavans de Ceccatty (1958) in *Hippospongia communis* and by Fell (1974), in *Haliclona ecbasis*, but in *Halisarca dujardini* (Lévi, 1965) and *Microciona microjoanna* (Fell, 1974) spermatozoa are randomly oriented. In view of the smallness of cells, and particularly of sperm cells, in sponges, microscopic observations have not absolutely excluded the possibility that nurse cells exist in some form, or that the germ cells are somehow ordered with respect to each other or to the follicular cells.

Present knowledge, although derived from an astonishingly small number of species, probably justifies the statement that spermatogenesis in sponges is basically similar to that of other Metazoa (Gatenby, 1919) and that its setting varies little throughout the phylum. However, the derivation of the germ cells appears to vary considerably even in closely related species, if not with respect to the cell of ultimate origin, at least with regard to the state of differentiation in which this cell is found when it transforms into a spermatogonium. Table 3 is designed to permit a quick survey of the main species investigated and illustrates the degree of variability of germ cell origin and the contradictions in the literature.

The spermatogonia of sponges have been studied hardly at all. Those investigators who considered them, surmised that they must undergo several, perhaps from one to four, divisions before they transform into spermatocytes. It is probable that in some species spermatogonia are interconnected. The majority of species show synchronicity of germ cell development within each follicle, but it appears well established that in a few species the germ cells even in one follicle develop asynchronously (for list, see Fell, 1974). It is of interest, that in some sponges all cysts in the body are developing synchronously, as reported for *Polymastia mammilaris* (Sará, 1961), *Axinella damicornis* and *A. verrucosa* (Siribelli, 1962) and *Tylodesma annexa* (Lévi, 1956).

One cannot escape the final conclusion that this phylum still invites intensive research on the reproductive process.

4

Cnidaria

Of the nearly 10,000 species of Cnidaria only the genus *Hydra*, one other athecate species *Eudendrium racemosum*, and the jellyfish *Phialidium gregarium* of the order Thecata, have been investigated sufficiently for a fairly coherent, detailed account of their spermatogenesis to be produced. Some pertinent data are also available for the order Limnomedusae. The information permits us to state that at least in the class of the Hydrozoa the process of spermatogenesis is more highly differentiated than in the Porifera, particularly in the way in which an ordered progression of stages of germ cell development is correlated with the behavior of certain associated somatic cells. There is some evidence that a similar pattern exists in sea anemones of the class Anthozoa. For the class Scyphozoa and for more than 99% of the whole phylum nothing can be said at present concerning specific details of spermatogenesis.

Spermatogenesis in *Hydra* the organism which has served experimental biology since Trembley's time (1744), has long been known to occur in testicles or "spermaria" which originate in the ectoderm from 'interstitial cells' (Kleinenberg, 1872). These organs have been studied again and again. The accounts of Aders (1903) and Downing (1905) particularly contain much detailed information, but we will begin here with the definitive descriptions given much later by Brien & Reniers-Decoen (1950, 1951) who have also briefly reviewed the pertinent literature. Fig. 9 tells a great part of their story.

In each testis of *Hydra* the germ cells are arranged in five or six layers each containing a generation. The youngest cells, the spermatogonia, lie at the mesogloeal lamella which separates ectoderm and entoderm and contains the spermatogonia; the oldest, the almost mature spermatids, lie under the epidermal surface. Throughout a spermarium the germ cells in each layer are in the same stages. The pithelial cells of the surface penetrate deep into the gonad with long processes subdividing the organ into compartments which Brien & Reniers-Decoen called 'follicles'. The walls of the follicles are incomplete and there appears to be no difficulty in channeling the spermatozoa from all parts of a testicle through the central opening of a "nipple" at the time of spermiation.

In *Hydra* gametes develop from interstitial cells whenever the proper, species-specific stimuli activate the process of differentiation. In *H. fusca* a stimulus, in this case low temperature, brings about such an outburst of gametic development, particularly in the male, that Brien (1966) has characterized it with the term 'crise

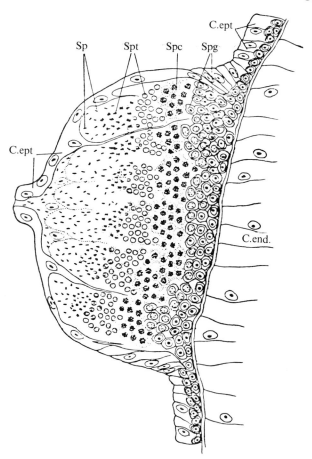

Fig. 9. Semi-diagrammatic representation of a functional spermarium of *Hydra attenuata*. Sagittal section. C. end., entodermal cells; C. ept., ectodermal, epidermal cells; Spg., spermatogonia; Spc., spermatocytes I; Spt., spermatids; Sp., "free" spermatozoa. The processes of the ectodermal epithelial cells run toward the mesogloeal lamella. (After P. Brien & M. Reniers-Decoen (1951). *Ann. Soc. R. Zool. Belg.* **82**, 285–327.)

gamétique'. The reservoir of interstitial cells becomes exhausted and the somatic tissues dwindle until the animal dies. In other species the stimulus for gameto-genesis is not so well known (Tardent, 1974). Spermatogenesis as in *H. attenuata* may not be restricted to particular seasons, nor do "crises" occur under natural conditions. As Brien has pointed out, there is no reason to assume that there is a separate group among the interstitial cells which can be conceived as a germ line ('lignée germinale').

According to Tannreuther (1908) each testicle of *Hydra* may spermiate several times within 40–50 hours, but Brien & Reniers-Decoen stated that a testicle is apparently capable of ejecting a generation of spermatozoa only once and undergoes involution after that. The involution begins with degeneration of primary spermatocytes. Young spermatids follow, while more mature spermatids are still seen moving about in apparently healthy condition within the reduced testicle. The spermatogonia seem to become reintegrated into the layer of interstitial cells at the mesogloeal lamella.

A somewhat different and more detailed description resulting from light and electron microscopic observation on *Hydra attenuata* has been given by Zihler (1972). He found that mature spermatozoa were released at intervals of 5–60 minutes in batches of 10–100. This agrees better with Tannreuther's than with Brien's observations. Zihler reported that different follicles were involved in the releases. He described that the surface epithelium opens at its thinnest point with each burst of spermatozoa, but he could not find the single "nipple" with a preformed canal which Brien & Reniers-Decoen saw in *Hydra viridis* and *H. attenuata*.

The cytology of developing sperm cells in *Hydra*, particularly gonia and cytes, is not sufficiently explored. Downing (1905) gave detailed descriptions of their differentiation, but the cells are small in all coelenterates and all light microscopic data must be regarded with some skepticism. For instance, Brien & Reniers-Decoen emphasized that spermatids always appeared as separate cells in light microscopic pictures, but Zihler found in his electron microscopic studies that in 10–20% of the maturation divisions the cytoplasm does not divide, so that the four resulting spermatids remain connected, or in Zihler's words 'within the membrane of the spermatocyte'. In a description of the ultrastructure of spermatogenic cells in *Hydra attenuata*, Stagni & Lucchi (1970) made no mention of incomplete spermatocyte divisions but found the spermatogonia connected by many intercellular bridges.

A marine athecate hydroid *Eudendrium racemosum* has been recently investigated with the light and electron microscope by Hanisch (1970). *Eudendrium* produces spermaria ('Hoden') on modified hydranths or 'blastostyles'. The spermaria form between ectoderm and entoderm from 'spermatoblasts', which are the offspring of interstitial cells found earlier in the entoderm. The term spermatogonia was not used by Hanisch. The masses of germ cells forming the tissue of the spermaria are at first delimited only against the ectoderm by a definite mesogloeal lamella which shows two dense marginal zones, one toward the testicle and one toward the ectoderm, sandwiching a less dense central layer. A second mesogloeal lamella forms later between entoderm and germinal tissue. This second lamella has a definite marginal zone only on the side of the entoderm, but remains ill defined at the side of the germ cells. It is perforated by pores of 0.1–0.4 μm through which extend processes of entodermal cells. Within the pores, in the entodermal cells and also extracellularly among the germ cells Hanisch observed dark masses, not bound by membranes, which he considered evidence that the entoderm conveys nutriment to the germ cells. No mention was made of any participation of epidermal cells in the organization of the gonads. When spermatoblasts become incorporated in a spermarium they decrease in size, lose their nucleoli and show a characteristic chromatin pattern. Cells of this type possessing a small flagellum are, according to Hanisch, primary spermatocytes. These become more frequent in older testes and show chromatin changes which indicate stages of meiosis. It appears that the last 'spermatoblast' division is often incomplete so that spermatocytes may be connected by intercellular bridges. Hanisch demonstrated quite conclusively that a flagellum begins to grow in the primary spermatocyte and that two flagella are present at opposite ends of the secondary spermatocyte, each spermatid receiving one in the second maturation division. He described the concomitant behaviour of the centrioles in detail.

In summary, while the process of spermatogenesis in *Eudendrium* is not known in all its phases, some outstanding features have been observed. The development of the germ cells proceeds from stem cells (interstitial cells?) to 'spermatoblasts' in the entoderm where the spermatoblasts begin to multiply. They finally become aggregated in spermaria and surrounded by mesogloeal lamellae and subsequently undergo meiosis and spermioteleosis. The qualitative and quantitative features of the spermatogonial stage are not well known. An unusual feature is the precocious development of flagella in the spermatocytic stage. In contrast to other hydrozoa no auxiliary cells have been observed in *Eudendrium*, but entodermal cells may play a direct role in the nutrition of germ cells.

While the accessibility of freshwater *Hydra* led naturally and early to investigations about its reproduction, the marine coelenterates do not lend themselves readily to being kept in the laboratory and have been used in experiments only in recent times. The first interest in medusae, for example, centered on their 'bilaminar', 'primitive' anatomy discussed by Thomas Huxley (1849). From this developed a strong interest in the morphology and embryology of organs, which in 1879 resulted in the publication of an exemplary paper by O. and R. Hertwig who described the gonads of a variety of hydromedusae as they appear in sections fixed in osmium tetroxide.

In the medusae, which are nearly always dioecious, the gonads are arranged along the radial gastrovascular canals or more centrally on the main gastrovascular cavity, and its outpouchings. They vary greatly in number (e.g. four in *Phialidium*, many in *Aequoria* or *Solmissus*). In cross-sections of the gonadal region the entodermal lining of stomach or radial canal is seen separated from the ectoderm by a thick mesogloeal lamella ('Stützlamelle') (Plate 2*b*). In the thickened ectoderm of the testes the Hertwigs recognized masses of developing gametes between the lamella and the thin epithelium of the surface. They distinguished cells of large diameter (9 μm) near the lamella and of small diameter (4.5 μm) under the surface, recognized that the relative frequency of cells of various sizes differed greatly in animals of different ages, and concluded that the smaller cells arise from the larger ones by division. In teased preparations they were able to show that the small cells may be in any stage from early spermatids to almost mature spermatozoa. The pictures and descriptions of the Hertwigs with regard to the relationship between epidermal and germ cells were not excelled until the advent of the electron microscope. The Hertwigs stated that the epidermal cells form a thin layer over the thick mass of germinal cells. The epithelial nuclei are regularly distributed under the surface, and cytoplasmic processes are penetrating into the germinal tissue at regular intervals. They begin near the surface with a funnel-like widening and end at the mesogloeal lamella with a small bulge. The Hertwigs considered these cells 'fused', i.e. a syncytium, but the electron microscope has corrected this view. The epithelial cells are separate entities (Roosen-Runge & Szollosi, 1965).

The investigations of the Hertwigs indicated that the pattern of spermatogenesis is probably very similar in all medusae although the gonads vary in anatomical location and degree of differentiation. The Hertwigs suggested a series of evolutionary stages based on the rule that 'an organ is the lower in the scale of evolution ('Entwicklungsstufe'), the less circumscribed and localized it is'. A species with definitely localized and highly differentiated gonads is the Leptomedusa *Phialidium gregarium* which has been investigated quite thoroughly by modern methods (Roosen-Runge, 1960, 1962, 1970; Roosen-Runge & Szollosi, 1965). Its spermatogenesis will be discussed at some length in the following. It represents one of the very few cases for which a partially quantitative account exists, and therefore forms an important prototype in the comparative considerations of this book.

Phialidium is a genus of small marine hydrozoans which is circumpolar in distribution in the Northern Hemisphere. Male and female medusae occur in large swarms and shed their gametes into the surrounding sea where fertilization takes place. In *Phialidium gregarium* sperm and eggs are shed nearly simultaneously twice a day, at sunrise and sundown. It has been established that the release is triggered by changes in light intensity from dark to light and from light to dark (Roosen-Runge, 1962). A single medusa if well fed may spawn quite regularly twice daily for the length of its reproductive life which is 3–4 months. The number of spermatozoa released by a male at each spermiation is of the order of 1½ million. All mature spermatozoa are released within a period of 15 minutes. Obviously a well controlled pattern of spermatogenesis must underlie such behavior, which is repeated at regular intervals for a long time.

The four testes are situated on the radial canals. They show essentially the structure described for various medusae by the Hertwigs, i.e. layers of germ cells in various stages stratified between a mesogloeal lamella and the epidermis (Plate 2a). The spermatogonia lie at the lamella and the most mature spermatids near the surface. The degree of reproductive activity as expressed by the size of the testes is directly dependent on the state of nutrition. In well fed animals the succession of stages and the arrangement of layers is very regular. The distribution of stages is not at all random but at least five distinct stages are found layered in each testis. The germ cells may be quantitated fairly easily and the kinetics of the cells can be derived by a numerical analysis of cell counts. From counts at different times of the day it appears that the turnover time of the germinal epithelium is approximately 3 days and that one complete generation of spermatozoa is released almost every morning and every night. This has been tentatively confirmed by labelling with [³H]thymidine (Roosen-Runge, unpublished data). However, the mode of entrance of spermatogonia (stem cells) into the cycle has not been established. Only in cases of gonad removal has it been shown clearly that new stem cells in the entoderm are activated to produce a new gonad (Roosen-Runge, 1965).

The epidermal cells penetrate the whole germinal epithelium with pillar-like cell bodies up to 200 μm long (Plate 2*a*). Each pillar has many fine lateral processes which come in contact with all germ cells (Plate 3) although they do not truly envelop them. At the surface of the testis the pillars flare out into ciliated plates which form a contiguous, tight epidermis. Whenever spermiation occurs the epidermis breaks open at the intercellular borders, the intercellular spaces of the superficial one-third of the gonad become open to the seawater and the mature spermatozoa escape in masses, partly propelled by their own flagella but mainly swept out by the current produced by the large cilia of the epithelial cells. The cells themselves, although their contiguity is interrupted at the surface, remain firmly anchored at the mesogloeal lamella and within minutes after the release of the spermatozoa the epithelium is restored to its former integrity. Germ cells of the immature generations are firmly held in the meshes of the supporting cells. They were never seen to float away during spermiation.

Additional data on coelenterate spermatogenesis are fragmentary. As far as the class of the Hydrozoa is concerned, the Limnomedusae are apparently very similar to the Athecata (*Hydra, Eudendrium*) and Thecata (*Phialidium*) as was shown clearly by Bouillon (1957), who also discussed briefly the older literature for the Limnomedusae. On the Siphonophora only the most rudimentary information appears to be available. The class of Scyphozoa has been rarely investigated. Aders (1903) described successive stages of developing 'nurse cells' ('Nährzellen') in the testes of the jellyfish *Aurelia aurita*. The cells develop in the entoderm and migrate through the mesogloeal lamella into the gonad. They are filled with granules and vacuoles and send processes between the germ cells. Their relationship to the spermatozoa seems particularly remarkable. Aders found the heads of the spermatozoa not only applied to, but often deeply invaginated into, the nurse cells. The life of the nurse cells always ended in degeneration and dissolution beginning in the immediate vicinity of the mass of spermatozoa.

Spermatogenesis in Anthozoa needs much more thorough investigation, particularly with the electron microscope. A short description and two excellent light micrographs were given for *Actinia equina* by Chia & Rostron (1970). Apparentley, the testes are tubular but their length and configuration remain uncertain because a three-dimensional analysis is lacking. On cross-section a tubule 'resembles the seminiferous tubule of a vertebrate'. The spermatogonia lie at the periphery against the basement membrane and the spermatozoa border the lumen with their tails hanging into it. The spermatocytes and spermatids occupy the intermediate zone. At certain stages the spermatids appear highly ordered in radial columns, an arrangement which may indicate the presence of some radially arranged supporting cells, but these have not been demonstrated. In another anthozoan, *Bunodosoma cavernata*, Dewel & Clark (1972) showed a very similar arrangement in what they called the 'testicular cysts', but their electron micrographs show no evidence of supporting cells. However, Dewel & Clark reported that the spermatids are connected by intercellular bridges, and it may be that the bridges cause the regular arrangement of the spermatids at certain stages. The authors also observed flagella and centrioles in spermatocytes which reminds one of Hanisch's (1972) similar findings in *Eudendrium* (see above). Lyke & Robson (1975) described the ultrastructural differentiation of spermatids in four species of Anthozoa. They found spermatids in groups of four connected by intercellular bridges. The electron micrographs show no evidence of any cells other than germ cells, although pieces of gonads were fixed.

For the related phylum of the Ctenophora no detailed account of spermatogenesis appears ever to have been published. In a recent ultrastructural study of *Beroe ovata* (Franc, 1973) spermatogonia, spermatocytes and spermatids are depicted, and it is stated that all cell divisions are incomplete and result in intercellular bridges. Certain electron-dense cells with long processes are described which separate groups of germ cells from each other. These do not resemble closely the supporting cells seen in Hydrozoa. There is no information on the dynamics of the process of spermatogenesis.

5

Mollusca

The reproductive tract of molluscs is often very complicated, but the anatomy of the gonads is essentially plain. Gonads are paired in aplacophorans, monoplacophorans, bivalves and a few polyplacophorans. Otherwise, they are fused into a single organ. Although the majority of molluscs are dioecious, hermaphroditism occurs frequently and prevails in pulmonates and opisthobranchs. Sperm and eggs may be produced side by side, or in separate compartments of an ovotestis, or in separate gonads. Gonads may occur as single long tubules as in the prosobranch *Acme fusca* (Creek, 1953) or as 'an irregular, branching and anastomosing system of canals' widely extending throughout the body, as in the California oyster (Coe, 1932). In opisthobranchs and pulmonates the gonads are usually composed of lobules (acini) in varying numbers. Wherever the histology of the gonads has been described in detail, a thin, cellular enveloping membrane is mentioned, which is usually 1–2 μm thick (Ancel's layer) and may have some connective tissue fibers applied to it. It was first described in pulmonates (Ancel, 1903) but was apparently found also in the primitive mollusc *Neopalina* (Lemcke & Wingstrand, 1959). Outside of this envelope there may be various unspecific layers of connective tissue, pigment cells, muscle cells, etc., but in general the gonad is contained in a very thin sac. A brief representative description for pulmonates is found in Abdul-Malek (1954b) or for mesogastropods in Sachwatkin (1920). The sizes of lobules are only rarely mentioned in the literature. They appear to range from 120–150 μm (*Littorina*, Linke, 1933) to 400–600 μm (*Vaginulus*, Lanza & Quattrini, 1964).

Spermatogenesis in molluscs, as in all invertebrate phyla, has been explored unsystematically and incompletely. Pertinent literature is quite voluminous, but it is confined largely to gastropods and a few bivalves, and even in these two large classes relatively few species have been investigated. For the classes of scaphopods and cephalopods information is so scanty that it cannot be used for comparisons.

Two special features of molluscan spermatogenesis, the hermaphroditic condition and the dimorphism of sperm, have attracted investigators from an early time, but these features have also caused the exploration of spermatogenesis *per se* to become unsystematic and the results to be confusing. Before the advent of the cell theory and before it had become established that spermatozoa originated from elements of the testis in all animals, Siebold (1836a, b) clearly distinguished two

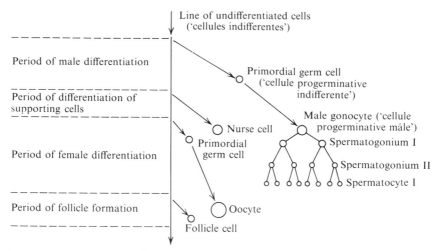

Fig. 10. Diagram of germ cell differentiation in *Helix*. (After P. Ancel (1903). *Arch. Biol.* **19**, 289–652.)

kinds of sperm in *Paludina*. His illustrations indicate that he observed many features pertinent to the process of spermatogenesis, but was unable to interpret them. He did not presume to know the final answer concerning the role of the sex organs in hermaphroditism but, contradicting Couvier, he stated his conviction that 'this much remains certain that in these ovaries of hermaphroditic snails sperm animalcules are found and may even develop there.'

Gastropoda
Pulmonata

There have been some significant contributions in the second half of the nineteenth century, a few of which will be mentioned below. Recent discussions, however, usually go back to the extensive, classical paper on the pulmonate snail *Helix pomatia* by Ancel (1903). His introduction begins: 'We have undertaken the study of a genital gland because in this field interesting questions of general importance and of great philosophic interest are particularly abundant.' The questions are those pertaining to the role and significance of the sexual products and to the mechanism of sex determination. Ancel's approach is a good example of the preoccupation of investigators with certain particular features of their experimental animals, to the detriment of research on the essentials of gametogenesis. He collected a large number of observations, surveyed the literature extensively, including even a comparative review of hermaphroditism in animals, but his account of spermatogenesis in the protandric vineyard snail is directed almost exclusively to the question of what determines the beginning of female development.

Ancel's view of germ cell differentiation in *Helix* is depicted in Fig. 10. In essence he concluded that male and female elements in the hermaphroditic gland develop from the same undifferentiated, progerminal cells which are induced to become either male or female by the environment, specifically by the absence in the case of the male, and the presence in the case of the female, of a 'nutritive material'. This he considered to be produced possibly by 'nutritive cells' derived from the same line of undifferentiated cells which give rise to germ cells. Male differentiation occurs first (protandry) in *Helix* as in most hermaphroditic molluscs. Indifferent progerminal cells differentiate into what we may now call male gonocytes, and they develop through a series of mitoses into primary and secondary spermatogonia, spermatocytes and finally into spermatozoa. When spermatogenesis is far advanced, some indifferent cells begin to differentiate into nurse cells distinguished by large size, characteristic chromatin configuration in the nucleus and by osmophilic granules in their cytoplasm. Ancel suggested that these cells are 'glandular' and secrete material which promotes, if it does not actually induce, the differentiation of female germ cells. He formulated a theory of 'accidental', i.e. not inherently determined, hermaphroditism which he considered applicable to many cases in the animal kingdom.

Ancel recognized no supporting cells specific for male elements. He identified two types of spermatogonia, primary and secondary, and considered the mitotic divisions of the spermatogonia unusual because they showed characteristic chromosome loops open toward the nuclear envelope. No detailed observations were reported on spermatogonial relationships, meiotic stages or spermiogenesis. This is all the more surprising as Platner (1885) had pointed out in *Helix* a rather intimate association of the developing spermatogonia and subsequent stages with certain cells of the gonadal wall: 'the spermatocytes become peculiarly related to certain cells of the alveolar wall which have transformed in a fashion to be described, in as much as all spermatocytes grouped around the same basal cell pass simultaneously through the subsequent developmental stages'. Platner had also described the regular arrangement of spermatid bundles around certain nuclei close to the acinar wall, similar to the observations of Siebold half a century before.

A very different description was given by Sóos (1910) for *Helix* (now *Arianta*) *arbustorum*. He observed early spermatogonia in association with certain 'base cells' which later were seen to become nurse cells of developing spermatids. Sóos regarded these nurse cells as 'equivalent to the Sertoli cells in the testes of mammals'.

Ancel's relatively simple concept of gonadal development was further contradicted in several aspects by Buresch (1912). This investigator found that the development of male germ cells in *Helix* (now *Arianta*) *arbustorum* is not followed by differentiation of nurse cells and subsequently of oocytes, but that all three cell types differentiate early. Oocytes do not mature until spermatogenesis has reached or passed its peak, but they are present and develop slowly during the male phase. Buresch was of the opinion that 'whether an indifferent sex cell will develop in male or female direction can be predicted early by the location of these cells close to or further removed from a nurse cell'. He considered it the decisive sign of maleness of a gonial cell that it becomes free of the wall, although it may remain close to it. Secondary spermatogonia he described as lying in groups and being 'grape-like', connected with each other. Spermatids are still congregated in groups and develop a 'cytophore' (see Glossary) which in Buresch's illustrations appears as an anuclear mass although it is labelled without further explanation as a 'degenerating nurse cell'.

Gatenby's (1917) investigation of still another closely related species, *Helix aspersa*, did not serve to clarify the contradictory evidence. Gatenby admired Ancel's descriptions but considered Ancel's fixation techniques inadequate. He also felt forced to discard the theory that yolk-producing nurse cells induce female development because he found them associated with male germ cells as well. In other respects Gatenby sided with Ancel. He found the same sequence of events from male to female development and saw cell borders in the germinal epithelium, which Buresch had not found. Finally, Gatenby made the point that in the ovotestis each new generation of sperm cells is subjected to a somewhat modified environment, although his conclusion that 'the nutrimental condition of the wall producing a male generation is hardly ever the same', is not based on good evidence.

In a subsequent paper Gatenby (1919) demonstrated in *Testacella* and *Helix* what he called 'giant germ-nurse cells', but this is certainly a misnomer. They are described as germ cells passing through all phases of meiosis up to pachytene when they die. They seem comparable to early stages of heterotypic spermatozoa rather than to nurse cells in the usual sense, even if, as Gatenby suggested, they release material which is utilized as food by surrounding germ cells (see Chapter 10).

The same species which Sóos and Buresch investigated was once more thoroughly explored by Eichhardt (1950). This author felt, and one can only agree, that the contradictions and incompleteness of previous results would justify a new start. She concurred with Buresch that the germinal epithelium is syncytial and does not contain cell borders. Nurse cells have at first no close association with oocytes (oogonial multiplication is said to be missing) nor with spermatogonia. The main phase of male development begins in the third month of development. The spermatogonia which have been lying loosely and irregularly in the lumen now connect with cytoplasmic strands extending from the germinal syncytium. In every case these strands project from an area where a nurse cell nucleus is evident. The spermatogonia assemble 'grape-like' around the nurse cell processes and establish 'syncytial' connections with them. Young spermatogonia are described as 'from the very beginning in connection with the germinal syncytium'. They move toward the lumen, ordered around the cytoplasmic strands projecting from the nurse cells. The spermatids remain attached to the nurse cells, but even after the spermatozoa have become detached they remain bundled. Eichhardt found nurse cells equally associated with oocytes and male elements and she rejected their role as determinants of sex.

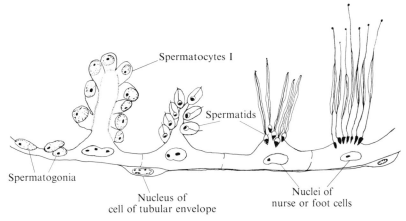

Fig. 11. Semi-diagrammatic representation of spermatogenesis in *Limnaea stagnalis*. (After V. E. Archie (1943). PhD Thesis, University of Wisconsin.)

Helix and its close relatives have supplied what is perhaps the largest group of directly comparable results in molluscs. Other pulmonates have contributed to the picture even if one cannot always be certain of the validity of the comparisons.

Crabb (1927) came to the conclusion that in *Lymnaea stagnalis* male and female germ cells as well as 'Sertoli' and egg nurse cells may arise simultaneously from indifferent epithelial cells and are commonly found developing together in a single acinus. He pointed to varied conditions in the same individual. Spermatids are sometimes seen to detach early from 'Sertoli cells'; in other instances the 'Sertoli cells' break loose from the walls with spermatids in advanced stages still attached.

Archie (1943) gave a less complex description of the ovotestis of *Lymnaea stagnalis*, which is semi-diagrammatically summarized in Fig. 11. The structure of the testis is similar to that of *Helix*. The germinal epithelium divides mitotically and appears to be composed of clearly demarcated cells. The spermatogonia develop in clusters (up to 20 cells) with their apices in close apposition and attached to a nurse cell ('Sertoli cell') by a cytoplasmic stalk. Archie assumed that a nurse cell and a spermatogonium are derived from a primordial cell in which the nucleus divides without cytoplasmic division. One nucleus gives rise to spermatogonia by further divisions. The other does not divide and becomes the nurse cell nucleus. The cells remain interconnected and confluent. A peculiar finding is the existence of secretory cells which appear only when spermatogenesis is already well advanced. These are large, bulb-shaped cells extending into the lumen by a cytoplasmic stalk and containing 60–120 round bodies which stain lightly and variably with iron-hematoxylin. The nuclei resemble those of nurse cells. They are polymorphic and possess a large (diameter 2.6 μm) nucleolus.

Aubry (1954*a*), in another investigation of *Lymnaea stagnalis*, concluded that nurse cells in pulmonates supply only male, not female cells and cannot be considered determinants of the female sex. She once again insisted on the syncytial character of the 'germinal epithelium'. Her paper ends with the statement: 'In their morphological characteristics, their ability to divide amitotically, their essential contact with male gametes and their potential phagocytic ability, the nurse cells in *Lymnaea* appear in every respect the homologues of the Sertoli cells in vertebrates', a concept which will be debated in Chapter 11.

The nature of the association between spermatogonia and nurse cells is not clear from Aubry's descriptions. The nurse cells apparently send processes into the lumen of the acinus and envelop the secondary spermatogonia which appear at this time (1954*a*). In another paper (1954*b*) the secondary spermatogonia are described in more detail. They are smaller than the primaries and occur in 'isogenic' groups. Clusters of synchronously developing germ cells attached to large nurse cells rich in alkaline phosphatase, with polymorphic nuclei, were described also for a slug *Philomycus carolinianus* (Kugler, 1965). In this case, the nurse cells become detached from the acinar wall while a cluster of 200 or more spermatozoa is still affixed to them. The spermatozoa are released as the whole assembly leaves the ovotestis, and the nurse cell then disintegrates. Watts (1952) has described observations on the living spermatogonia of *Arion subfuscus* as they develop in the germinal epithelium and subsequently pull out into the lumen.

Luchtel (1972*a, b*) investigated the gonadal development in *Arion circumspectus*, *A. ater rufus* and *Derocerus reticulum* with the electron microscope. He suggested that sex determination is due to the distribution of indifferent primordial germ cells into a medullar male compartment and a cortical female compartment, so that spermatogonia and oogonia develop in physiologically different environments. The undifferentiated auxiliary cells give rise to supporting cells (called 'Sertoli cells' by Luchtel) in the male and to follicle cells in the female part of the gonad. These cells apparently are consistently interposed between male and female gametes. Close junctions connect the 'Sertoli cells', and are also occasionally found between these and spermatogonia. The spermatocytes show no such junctions but often show cytoplasmic processes indenting the 'Sertoli cells'.

In *Gundlachia* the nurse cells are very similar to those of *Lymnaea* (Wautier, Hernandez & Richardot, 1966) and remain in association with surrounding cells. Subsequent development of oocytes may result in follicles which appear to carry clusters of spermatocytes. It is of interest that in this species the nurse cells do not appear to grow and spermatocytes remain small and comparatively few in number. Pelluet & Watts (1951) have stated that the development of nurse cells in the slug is accompanied by an increase in size of the mitochondria. During the development of the attached spermatids the mitochondria of the nurse cells decrease in size and number.

In a paper primarily concerned with the numerical composition of isogenic groups of spermatogonia in *Vaginulus*, Quattrini & Lanza (1964*b*) have published an electron micrograph which may be at present the only evidence of the nature of such cellular associations. The picture shows spermatogonia connected at their apices by cytoplasmic bridges. They are assembled around a process from a nurse cell. Although the resolution is not high, it is possible to identify cell membranes separating the cytoplasm of nurse cell and spermatogonia. This bit of evidence contradicts the concept of confluency as expressed, for instance, by Archie.

In one of the altogether rare explorations of living gonads, Merton (1924) found in *Planorbis* that 'basal cells' are associated with germ cells of both sexes throughout the whole of oogenesis and spermatogenesis. He demonstrated that they move in ameboid motion and are also phagocytic. The pictures of Buresch which showed spermatids not connected to basal cells, Merton rejected as artifacts. He also discovered a new phenomenon. He observed that the living basal cells release droplets ('Kinoplasmakugeln') which become associated with the sperm (Plate 5a) and then glide along its tail which is 700 μm long. These droplets of basal cell origin are not to be confused with parts of the cytoplasm of the spermatids, which have been shed before the droplets appear. Merton (1926) developed the hypothesis that the droplets of 'Kinoplasma' secreted by the basal cells and subsequently covering the surface of the spermatid, carry an agent ('Betriebsstoff') necessary for the motility of the spermatozoon. He surmised that the active compound is a lipoprotein organized for motility in the ameboid basal cells which synthesize it (myosin-like?). He rejected the concept that this substance plays a role in nutrition of the spermatid. In a later paper (1930) Merton expanded his observations. The following is a brief synopsis of his description of spermatogenesis.

In *Planorbis* the spermatogonia are situated at the proximal edge of each acinus where it opens into the atrium of the duct system. At the 'bottom' or the apex of each acinus the most mature stages of the spermatids are found, their tails extending through the lumen into the atrium (Plate 5a). Along the acinar wall the stages of spermatogenesis are sequentially ordered from the youngest to the oldest. This order is the result of a continual migration of basal cells which function as carriers. Spermatogonia form clusters and appear to be connected to the basal cell by a stem (Fig. 12). (Merton considered this connection syncytial.) Only during the second meiotic division does this association become less close, perhaps because of the rapid cellular movements which occur. In the spermatid stage the tight connection seems to be restored completely. The basal cells 'secrete' very large cytoplasmic droplets (up to 9 μm in diameter) which attach to the anterior end of the sperm tail. In dark-field they can be seen to supply the tail with a thin coating as they glide along it. The direction of their movement may be reversed by suspending the sperm in *Planorbis* blood or better in blood diluted with invertebrate Ringer's solution.

The role of basal cells as germ cell carriers in the process of spermatogenesis was observed in another member of the Planorbidae by Abdul-Malek (1954b), and in two species of doubtful systematic relationship (Veronicellidae) by Lanza &

Fig. 12. A clone of spermatogonia connected to the process of a basal cell of the snail *Planorbis*. (After H. Merton (1930). *Z. Zellforsch. mikrosk. Anat.* **10**, 527–51.)

Quattrini (1964). Quattrini & Lanza (1964*a*) have confirmed Merton's observations on the 'secretory' function of basal cells and added ultrastructural detail. In *Vaginulus* and *Laevicaulis* they found highly differentiated basal cells which grow to a maximum length of 50 μm. Spermatids are first associated with the periphery of these cells but later invaginate into the region of the large, polymorphic nucleus. The electron microscope reveals a peculiar mode of segregation of part of the cytoplasm of the basal cells. A cytoplasmic area becomes set aside ('individualiz-zarsi') and surrounded by a membrane. At first, it appears like a large cytoplasmic inclusion. How this inclusion is secreted as a droplet is not described, but at a later stage it is found attached to the head of the spermatid to which it adheres closely. It then migrates toward and along the tail. As the droplets are released from the basal cells, they vary greatly in size and shape, but during their migration their size decreases and they become spheres of a diameter between 4 and 5 μm. They contain mitochondria and microvesicles but no typical endoplasmic reticulum. At the time of release the entire basal cell appears to be histochemically devoid of RNA. After the droplets are released the remaining, nucleated part of the basal cell slowly degenerates and finally disappears.

The findings of Merton and of Quattrini & Lanza appear to present good evidence that a 'nurse cell' may transfer cytoplasmic material directly to a spermatid. Such evidence is still unique in animal spermatogenesis. Further studies with more adequate resolution are needed to clarify many important details concerning the transfer and the association of cytoplasms.

Serra & Koshman (1967*a*), in an almost unique study of details in spermatogonial development, found in *Cepaea nemoralis*, a pulmonate snail, four successive synchronous spermatogonial divisions resulting in 16 spermatocytes which entered meiosis simultaneously. These divisions presented only three different mitotic phases, prophase, metaphase and anaphase in the initial cell, and only metaphase and anaphase in the three following divisions. Serra & Koshman concluded from their observations that the first spermatogonium accomplishes all the synthesis of DNA for four cycles. These unusual observations need confirmation.

In another paper, Serra & Koshman (1967b) reported that the male nurse cells in *Cepaea nemoralis* and *Helix aspersa* arise through a somatic reduction division as haploid cells and subsequently differentiate as the chromosomes replicate endomitotically with a concomitant increase in nuclear volume. The nurse cells finally reach a very large size with nuclear volumes indicating extreme polyploidy (perhaps four times octoploid). The details of this finding are also unique although the polyploid nature of nurse cells has been suggested by others too.

In summary, the major papers on pulmonate spermatogenesis, and a number of less extensive reports, appear to indicate that there are at least two different modes of the spermatogenic process. In both, nurse cells play a prominent role. The first is examplified by *Helix*, *Lymnaea*, *Succinea* (Hickman, 1931); *Physa* (Tuzet & Marriaggi, 1951; Duncan, 1958); *Polygyra* (Pennypacker, 1930) and *Philomycus* (Kugler, 1965). In these species spermatogonia differentiate in an even distribution throughout the germinal tissue of the acinus, and the developing germ cells attached to nurse cells remain essentially stationary until the spermatozoa are released. The second, described for *Planorbis* (Merton, 1924, 1930), *Helisoma* (Abdul-Malek, 1954b) and *Vaginulus* and *Laevicaulis* (Lanza & Quattrini, 1964), is easily identifiable by the sequential order of spermatogenic stages along the wall of the acinus. In this type the basal cells carry the developing germ cells from the distal to the proximal end of the lobule where they are released into the lumen.

It is possible that nurse cells are absent in certain species. In *Vallonia* (Whitney, 1941) they appear to be at least very inconspicuous. Spermatogonia appear at first to be attached to the germinal epithelium by a pedicel, but later are described as occurring in clusters in the lumen. Spermatocytes may again be attached. Whitney suggested that cytoplasmic fragments discarded from the spermatids may contribute to the nourishment of all germ cells.

Prosobranchia

The pulmonates have been discussed first because of the relatively large amount of information available for this subclass of the gastropods. Another subclass, the prosobranchs, and in particular the order of the mesogastropods, has been explored quite frequently although only very few detailed studies exist. In prosobranchs hermaphroditism is not as predominant as in pulmonates, but dimorphism of sperm occurs very frequently and has attracted much attention (Chapter 10). This feature has dominated the approach to the study of spermatogenesis in most of the group. As a consequence it is very difficult to compose from the literature a simple and consistent account of sperm development in prosobranchs. An epithelial lining of the testicular acini has been reported repeatedly. Linke (1933) described in *Littorina* epithelial cells which form a continuous cuboidal layer after spermatogenesis has ceased, and phagocytize what is left of spermatozoa and 'nurse cells'. Creek (1951) emphasized that in *Pomatia elegans* the epithelium is different and consists of mucous cells filled with granules of secretion. The spermatozoa were seen with their heads embedded in these cells which apparently can also become phagocytic. In *Acme fusca* (Creek, 1953) the epithelium is composed of elongated gland cells between which lie groups of developing germ cells.

The "germinal epithelium" is generally assumed to be syncytial in character (Linke, 1933). Meves (1903) stated that in *Paludina* the layer of basal cells is syncytial and that the spermatogonia either develop within it or pull out into the lumen. I found no description of the early differentiation of spermatogonia. At a somewhat advanced stage of development, spermatogonia appear typically to lie "free" in the lumen, often in clusters (Ankel, 1930; Linke, 1933; Bulnheim, 1962a). There are other cases, however, in which spermatogonia are attached to nurse cells (Schitz, 1920; Furrow, 1935); or are described as in close association with the wall or near 'basal cell' nuclei (Sachwatkin, 1920). Ankel (1930) observed that in *Janthina* the typical gonia are found in the cavity even at the earliest stages, but the atypical ones which give rise to most unusual atypical spermatozoa (Chapter 10) remain connected with the wall by long cytoplasmic processes until they have almost reached the size of typical fully grown spermatocytes. Nurse cells have been described frequently, but it is probable that the descriptions refer to at least two different cell types. The first type is similar to the nurse cells described for pulmonates and often called 'Sertoli cells' (Meves, 1903). In *Ampullaria* (Sachwatkin, 1920) these are very large and contain polymorphic nuclei up to 20 μm long. Similar cells are found in *Turitella* (Schitz, 1920), *Valvata* (Furrow, 1935), and perhaps in *Melania* (Tokeya, 1924) and in *Murex* (Battaglia, 1954). The other type was probably first described by Reinke (1913) in *Littorina* and more extensively by Linke (1933) in the same genus. These cells occur in groups ('Nährzellen'). Their nucleus is relatively large and their cytoplasm vacuolated. Initially, they are connected to the wall by a stalk. Later, they are released into the lumen. 'Yolk granules' occur and gradually fill the cell. The nucleus becomes 'degenerate' at this time, but does not seem to disappear. Spermatozoa attach to these nurse cells and are found on them in the seminal receptacle of the female reproductive system. Cells filled with yolk granules or platelets and serving as free-floating carriers of spermatozoa during the last part of their maturation are certainly not similar to the nurse cells commonly described in many phyla and do not resemble Sertoli cells. Their characteristics are closely related to those of atypical spermatozoa discussed in Chapter 10.

A detailed account of the ultrastructure of a nurse cell exists for the mesogastropod snail *Cipangopaludina* (Yasuzumi, Tanaka & Tezuka, 1960; Yasuzumi, 1962). The origin of the association between nurse cell and spermatogonia was not investigated, but evidence was presented of a very intimate association of nurse cell and spermatids. Elongate processes protrude from the nurse cell toward the lumen and closely adhere to the head and middle piece of the typical spermatid. The processes become more slender and more numerous, approach each other and finally coalesce into a continuous thin sheet which envelops each individual spermatid in a mantle. The atypical spermatid has a different relationship with the nurse cell. It becomes lodged in a deep invagination of the nurse cell similar to that found in vertebrate Sertoli cells. Because of the great rarity of investigations concerning the ultrastructural details of these cellular associations, it is of special interest that Yasuzumi *et al.* found tubular extensions from the endoplasmic reticulum of the nurse cell extending in a very regular manner into the thin mantle which envelops the spermatid (Plate 5*b*). They suggested that the mantle was thus supplied with a canalicular system which may have a role in 'conducting nutritive substance from the nutritive cell to the developing spermatids'. The configuration is reminiscent of that in mammals where the acrosomal region of the spermatids is covered by a specialized marginal layer of Sertoli cytoplasm delimited from the bulk of the Sertoli cell by cisternae of endoplasmic reticulum (Brökelmann, 1963).

A histological and ultrastructural investigation of a dog-whelk *Nucella lapillus* was carried out by Walker & MacGregor (1968). The testis of this animal consists of numerous tubules which join to form a testicular duct. In the tubules the spermatogonia lie in groups around the periphery, mature spermatozoa are arranged around the lumen with their tails directed toward the center, and groups of spermatocytes and spermatids are scattered in between. Cells within the groups develop synchronously. Spermatogonia, spermatocytes and spermatids are all small and of virtually the same size, 3–4 μm in diameter. Different generations of spermatogonia were not identified, and Walker & MacGregor did not mention the presence of nurse cells although their electron micrographs do not exclude their existence. The filamentous spermatozoa, 80 μm in length, are remarkable because they are motile throughout their whole length. A flagellum runs from the front end through the center of the long nucleus to the tip of the tail.

Opisthobranchia

With regard to spermatogenesis, the opisthobranchs are the least investigated subclass of the gastropods. The few detailed descriptions which exist are relatively unremarkable. Hsiao (1939) found in *Limacina* that spermatids are 'arranged around a large degenerating cell which may be homologized with a "nurse cell" reported in most molluscs, but no cytoplasmic connection between it and the spermatids has been observed'. The spermatids later move away from the 'nurse cell', but spermatozoa, although not connected, still lie side by side oriented in the same direction according to Hsiao. Baba (1937), in an extensive study of the anatomy and histology of the nudibranch *Okadeia elegans*, briefly described the gonads. He mentioned that the spermatozoa, 'fastened together by their heads', come into close contact with the epithelial lining which he assumed to have a nourishing function. Incidentally, it is of interest that in this species the ovaries appear to arise at the surface of the testis. Usually the animal possesses two testes and five ovaries, and the mature eggs travel through the testes. The embryological condition may represent another case in which, as in vertebrates, the male gonads develop from a medullary, the female ones from a cortical part of the gonadal rudiment. Other cases have been described in molluscs, for instance by Furrow (1935; the paper contains a short review of the gonadal embryology in molluscs) and Luchtel (1972*a*, *b*).

The testis of an aberrant gastropod *Comenteroxenos* (Tikasingh, 1962) may on further investigation reveal some remarkable features. Tikasingh described the organ as ductless. The spermatozoa lie in the lumen of the single organ without any orientation, and it is entirely unclear how they are released.

Bivalvia

In classes other than gastropods spermatogenesis has been only scantily investigated. Some information exists on a small number of species of bivalves. The California oyster (Coe, 1932) presents some interesting features although the histology has been only superficially explored. According to Coe, the mature gonads in the oyster extend through almost the whole body as a branching and anastomosing system of tubules. Long before the whole system is mature, differentiation proceeds in limited areas.

It is frequently impossible to assign a sex to a young animal, because all possible mixtures of male and female development are found, but in older animals the sequence of phases occurs with greater precision. In the first year there are usually three well defined phases, male, female and male, in succession. Spermatogonia are situated within the wall in small synchronous groups intermingling with oocytes. Groups of large spermatocytes and small spermatids in various stages are seen in the lumen. As spermatogenesis proceeds the genital ducts become crowded with 'spermballs' (also called 'spermatophores' by Coe) containing 250–2000 spermatozoa derived presumably from a single spermatogonium. If this assumption is correct, six to nine spermatogonial divisions must take place. In a 'spermball' the heads of the spermatozoa are all turned toward the center and the ball is firmly cohesive, but no mechanism for the cohesion was suggested by Coe. The balls should be not called "spermatophores" nor is it probable that they are "cytophores", but they may represent bundles of spermatozoa still interconnected by cytoplasmic bridges or simply adhering to each other as in the spermatozeugmata of teleosts.

The papers of Weissensee (1916) on *Anodonta*, Woods (1931) on *Sphaerium*, Loosanoff (1937) and Ansell (1961) on *Venus*, indicate no unusual features of spermatogenesis and suggest that nurse cells, if they are present, must be quite inconspicuous. It appears, however, that in many species of bivalves the testicular follicles are surrounded by large cells which often form a many-layered envelope (Fig. 13). Loosanoff (1937) considered these to be 'nutritive' cells. Coe (1943)

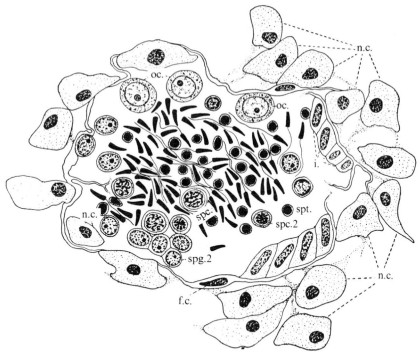

Fig. 13. Gonadal tubule of a young clam, *Venus mercenaria*. f.c., follicle cells; n.c., 'nutritive phagocytic cells' surrounding tubule; oc., oocyte; spg. 2, secondary spermatogonia; spc. 1 and 2, primary and secondary spermatocytes; spt., spermatid. (From V. L. Loosanoff (1937). *Biol. Bull.* **72**, 389–405.)

demonstrated that in *Teredo*, *Mya*, *Ostrea*, *Pecten*, *Mytilus*, *Anomia* and other species the follicles are surrounded by such presumably nutritive tissues, sometimes in large aggregates. In *Anomia*, for instance, these are assimilated during late stages of gametogenesis. Coe did, however, also describe the existence of small 'follicle' cells scattered among the spermatogonia which, at least in *Teredo*, undergo cytolysis and disintegration while the spermatogenic cells multiply. Their collapse results in the formation of a lumen in which spermatogenesis is completed, but their function is not clear.

Cephalopoda

I have not been able to find any comprehensive accounts of the testicular histology and spermatogenesis of cephalopods. The order Tetrabranchia appears not to have been closely investigated at all in this respect, and for the order Dibranchia only the classical papers by Pictet (1891) and Thesing (1904) have contributed substantially to the issue. I will limit myself to a brief review of Thesing's investigation, which in several respects corrects dubious earlier findings.

The major part of Thesing's descriptive work is concerned with details of spermiogenesis in *Octopus defilippi* and *Scaeurgus tetracirrus*. He did not specify the reasons why he found *Loligo* and *Sepia* 'unfavourable' for his research. From a great number of observations Thesing came to the conclusion that in the spermatogenesis of the cuttlefish *Rossia* a certain number of germ cells disintegrate at almost any stage, as spermatogonia, spermatocytes, spermatids or even as spermatozoa. In the testicular 'follicles' these form large masses into which the normal spermatozoa penetrate. Often, but not in the majority of cases, spermatozoa may be seen attached to somewhat smaller bodies which contain recognizable remains of nuclei and may give the impression of "cytophores" (Plate 6*f*). Pictet (1891) was under the impression that such structures do represent large nurse cells which form typical cytophores, but Thesing considered them as just one mode of formation of a nutritional mass. As the phenomenon has never been reinvestigated, it is difficult to assess its meaning. In Thesing's view male germ cells degenerate (up to 10% of them in *Rossia*) and form nutritive aggregates in those species among the cephalopods in which the spermatids have only a small amount of cytoplasm (*Rossia*, *Loligo* and *Sepia*). The process is of minor significance in species with large spermatids (*Octopus* and *Scaeurgus*). A comprehensive investigation of cephalopod spermatogenesis is urgently needed.

6.1

Arthropoda: Crustacea, Arachnida, etc.

The phylum of the Arthropoda represents three-quarters of existing animal species, and these show a vast diversity in their modes of reproduction. In only a very few species, most of them insects, which are treated in the second part of this chapter, has the process of spermatogenesis been investigated thoroughly. Whole classes have never been examined. If one excludes the insects, the remaining arthropods are still more numerous than any other phylum except the molluscs, and very atypical forms of spermatozoa and of reproductive behavior have been found among them, especially in the Decapoda and certain Cirripedia. Despite the interest which these have aroused, investigations have, with very few exceptions, concentrated on specific features of spermiogenesis and meiosis. For the purposes of this review it is the exceptional accounts which serve to elucidate the general process of spermatogenesis, and it is these few which will receive the greatest attention in the following. The picture which emerges is unusually incomplete even in a field where incompleteness is the rule, and it may not even be representative. If one judges by the diversity of detail indicated in the few cases of which we have any knowledge, it is in the Arthropoda where one should expect many and significant findings in the future.

The presentation in the first part of this chapter follows the systematic progression as indicated by Rothschild (1965), and begins with Diplopoda and Chilopoda, skips the Insecta, and continues with Crustacea and Arachnida.

West (1953) has given a lucid presentation of the microanatomy of the male gonads in a millipede, *Scyntonotus virginicus*. In this animal there are 36 'testicular vesicles', two pairs in each segment from the ninth to the seventeenth. Each vesicle contains only one generation of synchronous germ cells. Each is surrounded by large epithelial cells outside of which a fibrous connective tissue sheath is found. Terminal sacs of tracheae penetrate through this to the epithelium. The whole is enveloped by a network of small striated muscle fibers. Development of the vesicles proceeds from anterior to posterior. Spermatozoa were first seen in the most anterior segment at the last moult. Vesicles in which germ cells are in a spermatogonial stage are approximately 22 μm in diameter. Mature vesicles containing spermatozoa measure 130–150 μm. The epithelium is very flat until the spermatid stage. It then becomes columnar and develops a 'cuticular sheath' on the outside. As the spermatozoa reach maturity the vesicles develop a lumen. At spermiation, contractions of the muscle layer occur which express the sperm. The vesicles then collapse and the epithelial cells degenerate.

What West called 'testicular vesicles' in the millipede are obviously the counterparts of 'cysts' in insects, i.e. individual clones of germ cells surrounded by an epithelium ('cyst cells') metabolically closely associated with spermatogenesis. However, the vesicles are individually covered by a typical testicular capsule and although they are not combined into a larger gonad, they represent secondary compartments and deserve the name "testicular vesicles" or even "testes" (see Chapter 11).

Tuzet & Manier (1953) investigated spermatogenesis in *Himantarium gabrielis*, a centipede. (Their paper contains a brief review of research on chilopods.) The story is very different from that described in millipedes. *Himantarium* possesses two testicles which consist of joined tubules. The tubules are surrounded by a thin limiting membrane. The small spermatogonia (11.5 μm in diameter) are situated peripherally, the very large spermatocytes (80 μm in diameter) more centrally. The spermatogonia pass through an unknown number of mitoses. The resulting spermatocytes I enter a period of remarkable growth 'similar to that of oocytes in the beginning of oogenesis'. During this growth the cytoplasm differentiates into two portions clearly delimited 'by a kind of membrane', a central, acidophilic portion surrounding the nucleus, and a voluminous peripheral one which stains less distinctly. Tuzet & Manier called the central part the 'active', the peripheral part the 'accessory' cytoplasm. The nucleus does not participate in the growth of the cell, but during the whole period it appears to emit chromidial granules which disperse throughout the 'active' cytoplasm. Finally the oval spermatocyte reaches a size of 60×90 μm. As a consequence of this enormous growth the testicular tubules become tightly packed with large cells compressed into various shapes. Typical leptotene and diplotene stages and bivalents were seen. The young spermatid is 100–120 μm long and 60 μm wide. The circumstances of the formation of a flagellum and an acrosome were described in detail by Tuzet & Manier. The mature spermatozoon is thread-like and a very large amount of cytoplasm is apparently eliminated during the last stages of spermiogenesis. The spermatozoa are grouped in bundles and finally appear rolled up five or six together into dense cells. The unique cytoplasmic differentiation of spermatocytes revealed in this study and seen also by others (see Tuzet & Manier) stands in urgent need of electron microscopic investigation.

Tuzet & Manier made no mention of accessory or nurse cells in Chilopoda. On the other hand, Tönniges (1902) concentrated almost exclusively on such cells and their relationship to germ cells in *Lithobius fornicatus*, a common European centipede. He developed the concept that spermatogonia develop in and at the expense of a syncytial mass of 'nurse cells' which fills the center of the testis in early stages of development, but regresses later. Part of the spermatogonial population then degenerates and serves as nutritive material for the remainder. Because of this imputed function Tönniges called the degenerative cells also 'nurse cells'. The story is not sufficiently well documented to deserve much attention,

Fig. 14. A diagrammatic representation of the ultrastructure of a 'mid-body' bridge between spermatogonia of *Notodromas monaca*. (Redrawn from D. L. Gupta (1964). PhD Thesis, Cambridge University.)

but it indicates additional aspects of chilopod spermatogenesis which are in need of thorough investigation.

Significant accounts of spermatogenesis in Crustacea exist for only six of 35 orders listed in Rothschild (1965). The ostracod order Podocopa and the copepod order Calanoidea have been the subject of an important but unpublished dissertation by Gupta (1964). This work based on electron microscopic as well as light microscopic observations is not readily accessible, and will, therefore, be treated here in detail with special permission of Dr Gupta who has kindly supplied a few of his electron micrographs (Plate 8).

The testicular tubes of small freshwater shrimps are particularly favorable objects for study because in them the germ cells are arranged in strict chronological sequence of generations. The classical division is into four zones: I, the zone of spermatogonia; II, the zone of spermatocytes I in stages of meiotic prophase; III, the zone of the meiotic divisions; and IV, the zone of maturing spermatids. *Notodromas monacha*, one of the species first investigated by Schmalz (1912), has four tubular testes connected to a vas deferens on each side of the body. Each testis is surrounded by a very thin basement membrane (25–100 μm) which is strongly PAS (periodic-acid-Schiff reaction) positive, basophilic, metachromatic, and consists of a network of fibrils. It forms the basis for a continuous cellular sheath which according to Gupta may consist of only four 'sustentacular' cells per testis. These cells are not syncytial but interconnected by junctional complexes of the zonula adherens type. The nuclei are large and highly polymorphic. The cytoplasm forms extensions often less than 50 μm wide between the germ cells, but the intimate relationships to germ cells vary from one zone of the testis to the other.

Spermatogonia are scarce in the adult. In the subadult animal there are up to 16 in the anterior, blind tip of the testis. At least two generations of spermatogonia occur and the secondary spermatogonia are connected by intercellular, 'mid-body' bridges (Fig. 14). In the spermatogonial zone peculiar gaps in the plasma membranes are seen bridged by a row of vesicles, between spermatogonia and between spermatogonia and sustentacular cells. Often, if not always, the spermatogonia are separated from contact with the spermatocytes of the next zone by an envelope of processes of the sustentacular cells. Spermatocytes are regularly interconnected by cytoplasmic bridges but their cell borders adjoining sustentacular cells were always found to be continuous and intact. Spermatids are initially interconnected also. As they separate they become individually completely enveloped by the network of the sustentacular cells, which probably act in phagocytizing degenerate spermatids. In view of Gupta's descriptions of the intricate relationships between germ cells and large ramifying supporting cells, it is not surprising that similar cells have sometimes been said to *contain* rather than surround the spermatids, e.g. in *Daphnia magna* (Bérand, 1974), but the morphological evidence that this occurs is not conclusive.

Gupta emphasized that the greatly ramified cellular system occupying all space between the germ cells bears the same relationship to the latter as the cells of Sertoli in the mammalian testis. He found in the sustentacular cells of the copepod *Diaptomus* the ultrastructural characteristics commonly associated with steroid-producing cells, i.e. tubular mitochondrial cristae, paucity of ribosomes, lipid droplets and a large amount of smooth, often tubular, endoplasmic reticulum. 'If a sex hormone is elaborated by the gonads of copepods the sustentacular cells are the most likely site of formation.' Gupta also drew attention to the fact that in ostracods at the apex and the base of the testes the sustentacular cells show finger-like processes often filled with lipid droplets. As this group of animals has a large fatbody deposit between gonads and intestine, it is conceivable that the sustencular cells draw on this deposit and may be true "nurse cells".

A few details about the remarkable nature of the spermatozoa of *Notodromas* and of *Diaptomus* may be added. The sperm cell of *Notodromas* is 1 mm long in a male of 1.2 mm length. The seminal receptacle of the female is often as little as one-tenth the length of the spermatozoon. Each receptacle is connected to a vagina by an extremely coiled, thin duct which permits only one single spermatozoon to pass at a time. In *Diaptomus* the spermatozoa are released from the testis in immature spherical shape. They become elongate in the vas deferens and are incorporated in a 'spermatophore' where two types of cells can be found, the 'fertilizing' and the 'swelling' spermatozoa. Up to this stage the spermatids show no differences in light or electron microscopic examinations, but from elaborate histochemical tests done by Gupta it appears that 'the formation of spermatids with protamine-rich chromatoid bodies cyclically alternates with the formation of spermatids free from chromatoid bodies'. Gupta suggested that this latter type, which is poor in arginine and whose nucleoproteins are predominantly lysine-rich histones, gives rise to the 'swelling' spermatozoon, and that in it the chromatoid body which is present in *all* early spermatids becomes dispersed. To me these suggestions appear to be useful premises for future investigations on the determinants of germ cells.

In the Rhizopoda, an order of parasitic, very aberrant crustaceans, a predominant part of spermatogenesis occurs in the female tract. Formerly these animals used to be considered hermaphroditic because certain individuals appeared to possess functional ovaries and testes. Close investigation, however, revealed that the so-called 'testes' occurred in animals of female type. Reinhard (1942) discovered that larval males (cyprids) deposit masses of germ cells in the female where they develop into functional spermatozoa within sacs which should be called 'seminal receptacles' or 'spermatothecae' (Ichikawa & Yagamuchi, 1958). When the germ cells are transferred they appear to be in spermatogonial stages. The seminal receptacles are lined with an epithelium which is in a hypertrophied and degenerating state at the time when the male cells arrive. When they begin their development the epithelium breaks down completely except for the basement membrane. Ishikawa & Yagamuchi emphasized that not all cyprid germ cells transform into spermatozoa. Some remain on the wall of the receptacle in the manner of supporting cells. The implication is that the male deposits in the female not only germ cells but also somatic elements. In the genus *Thompsonia* Yamigamachi & Fujimaki (1967) demonstrated cells with very large nuclei (up to 60 μm in diameter) among the spermatogonia which they regarded as 'nurse cells' and which are presumably also derived from the male. The results of some experiments suggest that the entrance of male cells into the seminal receptacle is a prerequisite for the sexual maturation of the juvenile female rhizopod.

Spermatogenesis in Isopoda has been repeatedly investigated. Gilson (1884, 1886, 1887) examined the testes of several hundred species of arthropods and among them found some isopods particularly suitable for study. In his first paper he was particularly concerned with insects and with the question of 'female nuclei' ('noyaux femelles'). This term (see Chapter 2, p. 18) he applied to nuclei of the 'spermatoblasts' surrounding bundles of spermatids. From the context it is clear that Gilson's spermatoblasts are accessory or nurse cells. Gilson expressly stated that he would not prejudge the nature of these cells or ascribe a 'sex' to them. In the isopod *Asellus aquaticus* he found the 'female nuclei' extremely large and associated with relatively small numbers of spermatids. This was excellently demonstrated in his next paper (1886). Some of Gilson's pictures are republished here (Plate 7) because they tell the story better than Gilson's often verbose text with its poorly defined terminology. In *Asellus*, in each of the testicular follicles or tubules which caudally open into a vas deferens a definite sequence of stages occurs. For instance, in Plate 7(c) the most cranial mass of cells (below in the picture) consists almost certainly of spermatocytes with characteristic nuclear patterns. The cells in the blind end of the tubule Gilson called 'metrocytes' and regarded them as forerunners of a subsequent generation of spermatids. They are not seen in the figure. Successive stages of spermiogenesis are depicted in Plate 7(a), (b) and (c). In spermatid stages (Plate 7c, above) the germ cells hang together in bundles of eight or more. The nurse cells with large granular nuclei are regularly spaced along the wall, each associated with a cluster of spermatids. Initially the spermatids are attached not at the tip of the head but below it (Plate 7b and e). As the tail lengthens the sperm heads seem to become more proximally associated with the nurse cell and loops of tails point toward the lumen (Plate 7f). This peculiar configuration is almost certainly connected with the atypical morphology of the spermatozoon of *Asellus* which was later described in detail by Retzius

(*a*)

(*b*)

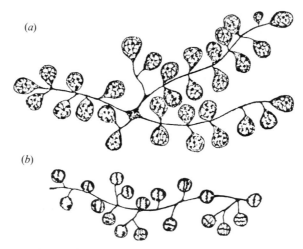

Fig. 15. Two cytophores of spermatocytes in the isopod *Portunion maenadis*, fixed and stained. (*a*) Spermatocytes I, (*b*) Spermatocytes II in metaphase. (After O. Tuzet & A. Veillet (1944). *Arch. Zool. exp. gén.* **84**, 1–9.)

(1909). According to Retzius the spermatozoon consists of a tubular part containing the head and an even longer stiff thread 'which one may call the tail'. These two parts are connected in such a way that they form a very steep angle with each other. In Plate 7(*f*) the junction between the two parts is presumably at the bend of the hairpin-shaped sperm. Retzius quoted Sars (1867) as having described that the spermatozoa gradually become free of the 'mother cell' which finally shrinks, and that they retain their strongly bent shape as they are transported to the outside.

Tuzet & Veillet (1944) described spermatogenesis in a parasitic isopod, *Portunion maenadis*. In this species the two testes are oval sacs which contain spermatogenic cells in synchronous groups. The groups are clones derived from single spermatogonia and are connected together by cytophores (see Fig. 15). Presumably there are five spermatogonial divisions because at the stage of primary spermatocytes 32 cells are interconnected in most cases, although the number is somewhat variable. Often one of the 32 cells is small and stunted. (The configuration is remarkably similar to one often seen in annelids, e.g. Chatton & Tuzet, 1941). The spermatocytes undergo considerable cytoplasmic and nuclear growth. Each cell divides perpendicularly to the axis of the cytophore. The spermatids are small (3 μm) and develop into non-motile spherical spermatozoa, which are in general characteristic for isopods, but of extreme smallness (2 μm). Tuzet & Veillet maintained that the cytophore breaks down after the meiotic divisions, but that the spermatids are nevertheless not developing singly, but in octuplets, 'rosettes', connected by protoplasmic processes which radiate toward a common center.

Reger (1966) investigated the developing spermatozoa of the isopod *Oniscus asellus* and the amphipod *Orchestoidea*, and found their ultrastructural morphology similar. Both have cross-striated 'tails' 400–500 μm long, but the tails do not have the structure of flagella and the mature spermatozoa appear non-motile.

Asellus apparently is particularly suitable as an object for investigations on the relationship between nurse and germ cells. Montalenti, Vitagliano & De Nicola (1950) have shown that the large 'follicular cells' depicted by Gilson contain pyronine-positive granules presumed to be RNA. The amount and the size of the granules waxes and wanes with the development of the germ cells. In the leptotene–zygotene stages and also in the diplotene stage of the primary spermatocytes, the granules reach peaks of their development. In the intervening pachytene they are small and sparsely distributed. During the maxima they are found also extracellularly, surrounding the spermatocytes. The authors believed that during these periods RNA is transferred, even though the transport could not be demonstrated morphologically. They considered the enrichment in RNA an essential step in the initiation of meiosis. Their findings were partly confirmed and extended in another isopod, *Anilocra*, by Montefoschi (1952) who showed that the secretory activity of the large polyploid nurse ('follicular') cells is related to the process of meiosis. Maxima of RNA staining occurred in 'secondary' spermatogonia after the last division (in other words in pre-leptotene primary spermatocytes) and later in leptotene and pachytene stages.

The Diplopoda commonly have tailless spermatozoa combined in couplets, 'binary spermatozoa'. Reger & Cooper (1968) suggested that the gametes in a couplet adhere by their concave surfaces in a specific manner, perhaps aided by a sticky, electron-dense substance produced in the spermatic duct. In another species Horstmann & Breucker (1969) described an enveloping acellular membrane ('spermatophore'). In still another species, Horstmann (1970) has described a 'phragmoplast' which appears in the mid region of the cytoplasmic bridges which connect early spermatids. It consists of continuous spindle fibers (microtubules), a regular network of endoplasmic reticulum and a spherical body of osmiophilic material, and somewhat resembles the 'midbody bridges' described in ostracods by Gupta (Fig. 14).

An amphipod *Orchestia gamarella*, has been studied by Meusy (1964, 1968) with the light and the electron microscope. In this animal there are two straight, tubular testes. The germinal zone where spermatogonia are generated extends as a longitudinal, narrow band from one end of the gonad to the other, similar to what has been described for decapods (see below). It consists of a germinal element, the gonia, and a somatic element, 'reticulum mesodermique'. The reticulum is not syncytial and shows no ultrastructural signs of synthesis or secretion, although it is very closely associated with the germ cells. Primary spermatogonia periodically separate from the germinal zone, divide, and become secondary spermatogonia. In electron micrographs the primary cells show fine, deeply penetrating intranuclear canaliculi. Their membrane is continuous with the inner lamina of the nuclear envelope. Chromatoid bodies are often situated where the canaliculi open. Secondary spermatogonia are similar. They show an increased RNA content, but Meusy considered this to be due to increased nuclear synthesis of non-particulate nuclear RNA and not to an increase in ribosomes. By means of labelling cells in DNA synthesis with [^{3}H]thymidine it was possible to determine the duration of the spermatogonial stage (7–8 days), the meiotic stage (7–8 days) and spermiogenesis (5–6 days).

A fairly extensive account of spermatogenesis in a stomatopodan species, *Squilla oratorica*, demonstrates many similarities between this order and the Amphipoda and Decapoda (Komai, 1920). The testes are tubular with a coiled distal end serving as a receptacle for the spherical spermatozoa. The germinal zone is a narrow longitudinal band in the mid dorsal line where a few large spermatogonia I and some 'nutritive cells' with polymorphic nuclei are situated. Spermatogonia and spermatocytes are intimately associated with nurse cells, but Komai did not discuss the relationship between spermatids and nurse cells.

The Decapoda with their atypical spermatozoa caught the interest of investigators at a very early time. Siebold described the peculiar non-motile sperm cells of the crayfish in 1836. Koelliker (1841) began his epoch-making dissertation on spermatogenesis with the 'Strahlenzellen' he saw in the efferent ducts of the lobster. Because these 'cells' did not move and did not resemble the typical flagellated 'sperm animalcules' or 'Samenfäden', he could not bring himself to declare that they were the essential elements of the semen (Siebold had been quite certain of that). On the other hand he became convinced that they originated in or from the round cells of the testis.

The first comprehensive report of the male reproductive organs of decapods was given by Grobben (1878). Although he did not clarify any cytological details, the paper is still of interest with regard to general anatomy. It contains, for instance, the information that in decapods with simple tubular testes only two-thirds of the wall is covered by germinal epithelium, while the remainder is occupied by a epithelial lining identical with that of the excretory duct. In more highly differentiated forms, e.g. *Astacus*, the testis may be branched and the testicular capsule continued in the form of partitions between 'acini' or coils of testicular tube. A germinal zone always exists as an anatomical feature. In Grobben's account there is also an intriguing reference to *Sida crystallina*, a species now classified as a branchiopod, in which five generatons of germ cells were found 'which had sprung in succession at regular intervals from the germinal zone'.

Binford (1913) studied spermatogenesis in the crab *Menippe mercenaria* and definitely established some of the outstanding characteristics of the process in decapods. *Menippe* has long, paired, tubular testes, 140–370 μm in diameter. They are lined with a thin epithelium which becomes thick and columnar in regions containing mature spermatozoa. A single row of nuclei which belong to stem cells or primary spermatogonia extends along the dorsal side of the testicular tubule. The secondary spermatogonia form a crescent in cross-section (Fig. 16). The germ

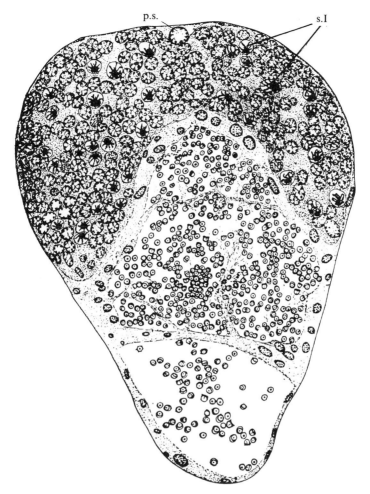

Fig. 16. Drawing of a cross-section of testicular tubule of the crab, *Menippe mercenaria*, ×450. p.s., "primary" spermatogonium; s.I., primary spermatocytes surrounded by secondary spermatogonia. Spermatozoa at bottom, spermatids above. (After R. Binford (1913). *J. Morph.* **24**, 142–204.)

cells develop in successive batches, and the lining cells form partitions between the generations. The spermatozoa which crowd toward one side of the tubule, reach maturity before the following generation attains the stage of synapsis, but they do not leave the testis until a third batch is under way or even a fourth one has started. As an interesting indication of the quantity of sperm production in this animal, Binford told of a crab which he observed spawning six times in 69 days during the summer. Each time ½ million to 1 million eggs were produced and all had been fertilized.

The papers of Fasten (1914, 1918, 1924) contain much cytological detail of the spermatogenesis of various decapods and a fairly extensive list of references. Cronin (1947) described the histology of the lobular testes of the crab *Callinectes sapidus*. Each lobule in this species contains several generations of germ cells. Cronin found accessory cells among the spermatocytes and, in reviewing the literature came to the conclusion that such cells exist in the testes of all decapods 'where histological description is adequate'. Matthews (1956) also found sustentacular cells among the primary spermatocytes in the testicular 'saccules' of *Hippa pacifica* and became convinced that they increased in number as the secondary spermatocytes increased, but both the origin and the means of multiplication of these cells remain obscure.

Word & Hobbs (1958) in a reinvestigation of the genus *Cambarus* discussed many of the problems which have presented themselves in decapods in general. In crayfish the testes are trilobed and each lobe is constructed in the manner of a compound acinar gland. The distribution of spermatogenic stages can be described by layers of acini. In the peripheral layer acini contain few or many spermatogonia, in the second layer meiotic divisions are seen, in the third spermatids are present, and in the fourth germ cells are in the last stages of spermioteleosis. The fifth layer contains some acini with mature spermatozoa in the center, others in which sperm is in the process of being extruded into the ducts, and some empty ones which are degenerating. Word & Hobbs described the life history of the acini in detail including the development of a sheath of connective tissue which in the tubular portions gradually acquires an outer layer of striated muscle fibers. The young acinus has no lumen. It contains very large spermatogonia (nuclear diameter 17.5 μm) and small accessory cells for which the authors created the new name of 'otheocytes'. The number of spermatogonia increases, but the accessory cells hardly multiply at all. Meiotic divisions occur synchronously in each acinus. Some tri- and tetrapolar divisions were seen which were out of phase. These may be divisions of 'otheocytes'. During spermiogenesis the spermatids appear to be embedded in a cytoplasmic mass without discernible cell borders. At this stage large, lobed nuclei, presumably of accessory cells, were observed at the periphery of the acinus. The central mass was considered to consist of an 'otheocytic syncytium' and of cytoplasm sloughed by the spermatids. While the spermatids mature within a gradually expanding acinar lumen, they are not arranged in bundles, but in rows between what appear to be territories of accessory cells. The whole story is full of detail which needs to be subjected to electron microscopic analysis.

It seems quite clear that each acinus of the crayfish testis serves only a single batch, perhaps only a single clone, of germ cells, and degenerates after spermiation. But the testis produces new batches in two or three successive spawning seasons. New acini must, therefore, be formed. Word & Hobbs assumed that these originate from the epithelial lining of the efferent ducts. The authors discussed also the gonadal transport of the non-motile spermatozoa and ascribed it to a combination of the 'pushing' action of the growing otheocytes within the acinus and the muscular action of the sheath of the efferent ductules.

Although the number of decapod species investigated is not sufficient to attempt a meaningful synopsis of the whole order, one gains the impression that the spermatogenic process in species with simple, tubular testes progresses in large synchronous groups more or less delimited from each other by perhaps temporary linings of somatic cells. In species with more complex testicular development, the batches or clones of germ cells become more thoroughly separated in acini or lobules.

Information on the process of spermatogenesis in the class Arachnida is dominated by the work of Sokolow (1912, 1926, 1929*a*, *b*, 1934) who studied various species of four orders, reviewed much of the scanty literary background, and supplied a large amount of directly comparable data. Among the arachnids spermatogenesis is best known in scorpions, pseudoscorpions, harvestmen, and mites and ticks. Only very few true spiders have been investigated and I have been unable to find any information on six of the eleven orders of the class.

The testes of scorpions (Sokolow, 1912; Abd-el-Wahab, 1957) are paired structures, each consisting of two longitudinal tubes which are cross-connected by usually four anastomoses into a ladder-like structure. According to Sokolow, in *Euscorpius* the wall is made up of tall columnar cells which are filled with osmiophilic granules. The testes are crowded with spermatogenic cysts ('follicles') and have little or no lumen. As in insects these consist of a wall of 'cyst cells' surrounding a synchronous aggregation of germ cells. The cysts are formed around spermatogonia. Cysts of different stages are intermingled. There is no zonation along the testicular tube. Initially the nuclei of the cyst cells resemble those of the testicular lining. Later they become folded and flattened. They may divide and frequently show polyploid metaphase plates. Spermatozoa are seen still in the cysts, rolled up in tight bundles. All these findings were confirmed essentially in *Buthus quinquestriatus* (Abd-el-Wahab, 1957).

A remarkable datum has been provided by Wilson (1916) for the scorpion *Centrurus exilicauda*. In this species, 'alone among animals hitherto examined, it is possible to conclude with certainty that the chondriosome material in primary spermatocytes is divided with exact equality among the spermatozoa'. Wilson described that a single ring-shaped mitochondrial mass in the spermatocyte is divided in such a way that each spermatid received exactly one quarter of it. By contrast, in a related species *Opisthocanthus elatus* Wilson found in the primary spermatocytes 24 large mitochondrial bodies scattered without order. The segregation of these bodies during the maturation divisions was inexact, 73% of the spermatids receiving six mitochondria, 16% five and 11% seven. Wilson concluded: 'It is evident that chondriosome material having the same origin, fate and (presumably) physiological significance may be distributed to the germ cells by processes widely different even in related animals.'

In pseudoscorpions, spermatogenesis is similar in many respects. In *Obisium muscorum* and *Chelanops cyrneus* (Sokolow, 1926) the testes consist of three longitudinal tubes, two dorsal and one ventral, interconnected by cross-anastomoses. All tubes have the same histological structure. When a primary spermatogonium has divided several times the neighboring 'germinal epithelium', i.e. somatic tissue, forms a thin covering for the group. In the mature animal this tissue has almost disappeared. As in scorpions, the cysts or follicles contain synchronous clones, but there is no zonation of stages in the testes. The most frequent stages are early spermatocytes and spermatids indicating, as Sokolow suggested, that these are of long duration. The spermatids are radially arranged in the cysts, often in whorls with the heads pointed toward the center. According to Sokolow they become individually 'encysted' by a surface modification of their own cytoplasm. Each spermatozoon is thus rolled up in its own capsule (the term 'cyst' is confusing here). In hypotonic solutions it unrolls and is set free. This presumably occurs also *in vivo* in the female reproductive tract. The capsules are normally combined into 'spermatophores' (see Weygoldt, 1970). In the pseudo-scorpion *Hysterochelifer meridianus*, Boissin (1970) considered the early testicular epithelium a syncytium in which somatic and germinal nuclei occur. Spermatogonia separate out of this epithelium and each of them through successive divisions gives rise to a cyst which is surrounded by somatic cells. As germ cells mature and cysts are emptied, new cysts are formed by the testicular epithelium. Boissin observed secretory products between the cells in cysts but also in the lumen of the testis between the cysts. She considered the role of the somatic 'syncytium' twofold, i.e. nourishing the germ cells directly and supplying a medium in which the cysts float. Boissin reported some rare data on cellular kinetics. In *Hysterochelifer* males live from 18–24 months and have two reproductive periods in successive Julys. The spermatogonial population is building up from August to May and then diminishes rapidly, staying at a minimum through the summer months. The relative number of cysts containing spermatocytes begins to increase in September and is at a maximum in December and January. Apparently the spermatocytes I rest through the winter months in meiotic prophase and do not proceed with their development until the spring. Many questions remain, particularly concerning the details of spermatogonial kinetics and the arrest of spermatocytes.

Spermatogenesis of Opiliones (harvestmen) has been investigated by Sokolow (1929*a*, *b*) in seven species, six of the same family. All are similar and the first and most thoroughly described, *Nemastoma lugubre*, can be regarded as the prototype (1929*a*), which is generally similar to scorpions, pseudoscorpions and spiders. The testis is a horseshoe-shaped tube across the axis of the body. At either end it is connected to a vas efferens. It possesses a peritoneal envelope reinforced by a weak tunica muscularis typical for the Arachnida. A 'germinal epithelium' from which germ cells and accessory cells develop is arranged at the periphery in groups of varying size. Primary spermatogonia are associated with these groups. Secondary spermatogonia are smaller and found more centrally in synchronous groups of two, four, eight or more cells surrounded by cyst cells. Sokolow postulated

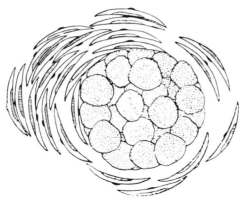

Fig. 17. Section of the content of a spermatocyst of the daddy longlegs *Nemastoma lugubre* with spermatozoa peripheral to a central accumulation of residual bodies (see text). (After I. Sokolow (1929*a*). *Z. Zellforsch. mikrosk. Anat.* **10**, 164–94.)

five to six spermatogonial generations in *Nemastoma*. Spermiogenesis is unusual because the spermatozoa in Opiliones are small concave–convex discs, aflagellar, and with extremely reduced cytoplasm. *Nemastoma* has the largest sperm (6.5×8 μm), in the other species measurements ranged from 1.75–3.5 μm. The spermatids begin as fairly large cells. During differentiation the greatest part of their cytoplasm is sloughed and the resulting residual bodies accumulate in the center of the cyst. The maturing spermatids remain marginal in random orientation (Fig. 17). When the cysts rupture, spermatids and degenerative cytoplasmic remains are released into the testicular lumen. Sokolow did not describe their subsequent behaviour.

There are no major accounts of spermatogenesis in the true spiders of the order Araneae. Koltzoff (1908) in a comparative study of 'head skeletons' ('Kopfskellette') of spermatozoa, described the spermatozoa of spiders as being encysted individually in hard shells which resist boiling in alkaline solutions and are presumably chitinous. Warren (1930) found that in the genus *Evarcha* the spermatids divide once or twice and give rise to two or four spermatozoa. He described the behavior of the chromosomes in these most unusual postmeiotic divisions, but his findings have been put into serious question by Sharma (1950) and others. Although I have found no thorough description of the whole spermatogenic process, there is little doubt that it is essentially similar to that in scorpions, pseudoscorpions and other arachnid orders as Sokolow has repeatedly pointed out (e.g. 1929*a*).

The order Acari includes the mites and the ticks. On the first, a truly comparative study exists (Sokolow, 1934) in which six species of the genus *Gamasus* were studied in considerable detail. In all species spermatogenesis occurs in cysts which are formed in the spermatogonial stage. The spermatogonia arise in a 'germinal center' ('Keimlager') on the dorsal aspect of the single voluminous testis. Some

TABLE 4. *Some features of spermatogenesis compared in six species of the acaran genus* Gamasus (*Sokolow, 1934*)

Species	Size of spermatozoon (μm)	Number of tids per cyst	Number of cyst cells per cyst
G. brevicornis	125–130	16	8
G. septemtrionalis	?	16	8
G. magnus	60–75	16	4
G. kraepelini	65–70	8	2
G. sp. no. 5	35–48	8	4
	Width 10	Sometimes 6	4
G. sp. no. 6	40–52	8	?
	Width 15		

data concerning the number of divisions of spermatogonia and cyst cells and the size of spermatozoa are collected in Table 4. Fig. 18 (*a*), (*b*), and (*c*) illustrates different ways in which the spermatids are arranged in closely related species. The spermatozoa in general carry much cytoplasm and are not only variable in shape and size, but also in the degree of condensation the chromatin undergoes in the nucleus. In one species (*G. magnus*) there is hardly any condensation. The spermatozoa are non-motile in the male tract but undergo much further differentiation as they become enclosed in a spermatophore and are thus transferred to the female. It is of special interest that the numerical relationship between cyst cells and germ cells is species-specific and shows little variability within the species.

Ticks have been studied extensively and in detail because their spermatozoa are very unusual in size, amount of cytoplasm and mode of motility. Tuzet & Millot (1937) began an article on spermiogenesis in the family Ixodides with the following statement: 'Rich as spermatogenesis appears in intricate processes and unique phenomena in numerous groups of animals, it is in the Ixodides where it presents, perhaps, the most complex features.' Tuzet & Millot reviewed the literature on the subject and described the growth of spermatocytes and spermatids as well as the acrosome and the small nucleus of spermatozoa in ticks, but information on the histology of the testis and clarification of the cellular structure of the unusually complex spermatids is better gleaned from more recent authors. The development of the 'giant spermatozoon' of *Argas columbarum* was studied by Oppermann (1935). The relatively immature cell which is released from the testis of this species is called a 'pro-spermium' and is more than 600 μm long. It continues to differentiate and grow in the female until it exceeds 1 mm in length. The size may be an adaptation to very long periods of storage which the sperm may undergo in the male and in the female. Spermiogenesis is dependent on a blood meal. Copulation occurs independent of the state of nutrition, but the final development of eggs and sperm in the female is again dependent on a blood meal. The large

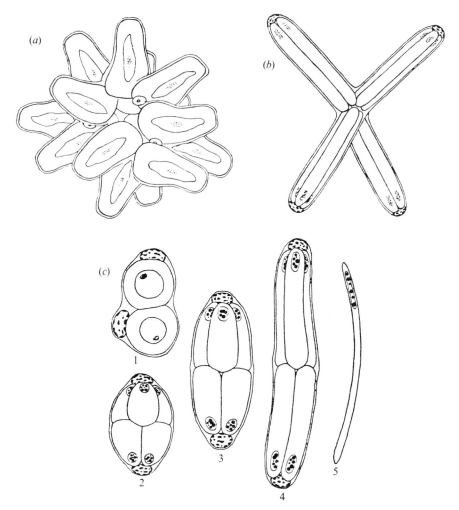

Fig. 18. Semi-diagrammatic drawings of spermatocysts in three different species of mites of the genus *Gamasus*. (*a*) *G. brevicornis*. A cyst with radially arranged spermatids (slightly misplaced through pressure on the cover glass). (*b*) *G. magnus*. Cyst with 16 spermatids. (*c*) *G. kraepelini*. 1, two spermatocytes I; 2–4 three successive stages of spermatid cysts; 5, mature spermatozoon. (After I. Sokolow (1934). *Z. Zellforsch. mikrosk. Anat.* **21**, 42–109.)

pro-spermium is in great part a carrier cell. The small nucleus lies at its posterior end from which a long tail process is formed containing a thin thread. According to Oppermann the testis has an unpaired tubular end where spermatogonia develop. This is continued in two long and coiled bilateral tubes in which all older stages are found. Cysts are conspicuous in the spermatocytic stage. The primary spermatocytes become very large (average diameter 88 μm) and their cytoplasmic interconnections show clearly in the light microscope. Between the cysts 'somatic supporting cells' are found which appear to degenerate in the stage of the secondary spermatocytes. The last reach more than 100 μm in diameter (it is unique that spermatocytes II grow larger than spermatocytes I) and Oppermann suggested that the degenerating supporting cells serve as nutriment for the germ cells. Spermatids and pro-spermia show no directional preferences in their development and grow into every available space.

The ultrastructure of the remarkable spermatozoa of ticks has been elucidated by Rothschild (1961), Reger (1962, 1963), and Breucker & Horstmann (1972). Rothschild experimented on the living spermatozoa and found that their motility is associated with long ribbon-like processes on the surface of the club-shaped sperm. He examined the living and fixed sperm with the light and the electron microscope. He found no flagellar structure but described longitudinal, parallel, ribbon-like processes on the surface of the cells, which are now known to be surrounded by modifications of the endoplasmic reticulum or the plasma membrane. Each process contains a single pair of microtubules, 6–9 nm in diameter (Reger, 1963), and it is suggested that the spermatozoa move by propagating waves along the processes.

Breucker & Horstmann (1972) have given an extensive account of the genesis of these peculiar spermatozoa in the tick *Ornithodorus moubata*. In the paired testes spermatogonia are found only in a narrow 'germinal zone'. In an initial proliferative period they undergo complete cell divisions and single spermatogonia are found surrounded by somatic cells. These are called spermatogonia A. After mitosis these cells remain interconnected and are then called spermatogonia B. Spermatocytes occur in groups of 16 indicating four successive divisions of each B clone. Spermatocytes become separated into clonal groups by septa. Breucker & Horstmann preferred to name the resulting compartments 'chambers' rather than 'cysts'. The chambers move caudad in large groups. In the most caudal area the septa finally dissolve, the spermatozoa reverse their route and leave through the germinal zone into the vesiculae seminales where they become incorporated into spermatophores. The final morphology of the spermatozoa is thus attained outside the testis. The primary spermatocytes undergo conspicuous growth. Their nucleus increases from 5 μm in diameter in the beginning to nearly 30 μm before diakinesis. Within a single chamber the nuclear diameters vary little. Intercellular bridges are prominent between spermatocytes. It is of great interest that the specific structures of the tick spermatozoon, the surface ridges, begin their complex development through invaginations of the plasma membrane early in the stage of the primary spermatocyte. In the tick *Amblyomma dissimili*, Reger (1962, 1963) described the surface modifications as arising from the endoplasmic reticulum. For the intricate details of spermiogenesis the reader is referred to the extensive accounts of the authors quoted above.

6.2

Arthropoda: Insecta

Insects constitute 90% of the phylum Arthropoda and nearly three-quarters of all animal species. The literature on insect spermatogenesis is extremely large, although here also this is caused by the great number of species available rather than by relatively broad coverage or unusually thorough investigations. Insect spermatogenesis provides the classical examples of clonal, encysted development of germ cells. In addition it demonstrates an accessory feature which is probably unique: the *apical cells*.

In this chapter, cyst development and apical cells will be treated extensively. The cytology of germ cells in all its huge variety will be discussed only very briefly. Some additional subjects such as cell kinetics and cellular degeneration will be treated in Chapters 9 and 10.

General reviews of testicular morphology in insects have been written by Deegener (1928) up to 1911 and by Depdolla (1928) up to 1926. Deegener presented a systematic survey with emphasis on gross morphology. Depdolla concentrated on the cytology of sperm development but described also many histological details. A highly competent and condensed summary of the structure of the male gonads may be found in Snodgrass (1935). The testes of insects usually are paired, but they are frequently united into a single, median organ. Testes consist of series of more or less parallel tubes, tubules or "follicles", emptying into a common vas deferens through narrow vasa efferentia. In some groups the testes are single tubules, often long and coiled (for instance, in *Dytiscus*; Demandt, 1912). In some Diptera each testis consists of only one follicle (Phillips, 1970). Measurements of follicles have been published infrequently, but from the few data available the order of magnitude is between ⅕ and ½ mm, i.e. the diameter of the follicles or cysts is of the same range as that of the seminiferous tubules of mammals.

Tunica propria of testis

In insects, the sheaths of the testes appear more closely connected with the process of spermatogenesis than, for instance, the tunica albuginea of vertebrates, and are, therefore, worth a certain amount of attention. They have been well described. Cholodkovsky (1905) distinguished in Lepidoptera and Diptera an outer, adventitial layer of varying thickness and an inner sheath, a membrane or tunica propria, which marks the outlines of the follicles. Tracheae are plentiful in the outer sheath and penetrate into the inner with fine ramifications, but never enter the cavity of a follicle.

According to Cholodkovsky the tunica propria consists of two layers, an outer epithelial stratum, often pigmented, and a translucent inner lining which on close inspection was found to be nucleated. From the inner layer fine cytoplasmic processes extend into the follicles. A more detailed and more dynamic description was given by Zick (1911) for two Lepidoptera. He considered elongate cells in the outer layer as pathways of nutrition for the inner one, and emphasized that the tunica grows and differentiates as spermatogenesis proceeds and flourishes. In early stages the follicle is filled tightly with cysts. Later the cysts appear to lie in a space filled with fluid which Zick assumed to be nutritional and secreted by the inner layers of the tunica propria. When spermatogenesis ceases the tunica dwindles and many of its degenerating cells may be extruded into the follicles.

Pigments are characteristic for the tunica and may color the insect testis bright yellow, orange, red, brown, greenish or violet. Zick found that pigmentation was absent in testes undernourished because of parasitic infections. Hadorn, Remensberger & Tobler (1964) determined that the testicular envelopes of *Drosophila* contain the highest concentration of isoxanthopterin in the larval body, although this compound is present also in the fat body, the intestine and elsewhere. If the larval abdomen is isolated by ligature and thereby prevented from undergoing metamorphosis, no isoxanthopterin is found in the investment of the testis, although its concentration in other places in the body is unchanged. Apparently the cells of the envelope attain the adult state of structural and chemical differentiation only when they are under the influence of an environment which undergoes metamorphosis. On the other hand, spermatogenesis proceeds with the same speed as in the controls. These results make it seem that the remarkable pigments in the tunica propria probably have no direct correlation with spermatogenesis, at least in its later phases, but they emphasize the specificity of the accumulation of these substances, presumably pteridines, which probably play an important role in processes of growth and differentiation (Gilmour, 1961).

The nutritive function of the tunica propria was confirmed by Buder (1917) who observed that a maximum of granules and droplets in the cells of the inner lining coincided with the time of maturation of the spermatids in the late larval and early pupal period of the butterfly. It remained uncertain whether the transfer of substances occurred directly from cell to cell or via the follicular fluid. In the last stages of spermatogenesis the sheath cells degenerated and products of their dissolution appeared.

Goldschmidt (1917) attempted to elucidate the physiological role of cells surrounding the follicles of lepidopteran testes. He maintained the follicles of the *Cecropia* moth in sterile hemolymph where the germ cells lived and developed for up to 3 weeks, the follicular and blood cells for 6 months. When the hemolymph was diluted, the follicles first swelled, and so did the individual germ cells in them, but this process was soon reversed and conditions in the follicle were gradually restored to normal. When germ cells alone were subjected to diluted hemolymph they swelled but never returned to their normal state. In fact, when the continuity of the follicular wall was even slightly disturbed, spermatogenesis failed to proceed. In older cultures, as soon as the germ cells began to degenerate, the follicular cells showed extraordinary activity, proliferating and growing out as epitheloid sheets. Goldschmidt concluded: 'The follicular membrane creates the specific osmotic conditions which at each stage of spermatogenesis force the cell to undergo certain changes'.

Goldschmidt further considered that the follicular walls might under certain circumstances mediate small physical or chemical changes within the follicle as a result of which the germ cells develop into oligopyrene or apyrene spermatozoa (see also Chapter 10 and Glossary). Cultures of follicles were always eupyrene in November and December, but in February, i.e. in the later pupal stages, only one in a thousand follicles developed and those which did were usually apyrene. Development could be made to proceed at higher temperatures but apyrene spermatozoa were still produced in large numbers. Goldschmidt suggested that the essential cause for these changes lies in the hemolymph. Knaben (1931) has cited against Goldschmidt's concept the fact that in the butterfly *Tischeria* normal and abnormal spermatids occur simultaneously even within a single cyst (Plate 10*b*). However, in this species spermatogenesis is normally not as synchronous as is usual in insects. Since Goldschmidt's time it has become abundantly clear that humoral stimuli exert very specific influences on spermatogenesis in insects.

As indications for a special physiological role of the testis sheaths become stronger, it becomes more interesting to investigate the morphology in still greater detail. Masner (1965) found in Hemiptera no morphological evidence that the tunica propria is instrumental in transferring materials from hemolymph to follicle. On the other hand, Wake (1963) observed with phase microscopy strange endomitoses and a state of extreme aneuploidy in the inner sheath of an orthopteran insect. Cantacuzéne (1968) showed a photograph of a polyploid parietal cell against which the heads of spermatids abutted.

In the first detailed electron microscopic study of the testicular envelope of an insect, Bairati (1967) showed that in *Drosophila* the wall consists of an external layer of pigment cells and an internal one of myoid cells which are on the average only 0.4 μm thick. The last contain two kinds of filaments, large (length 15 nm) and small (length 3–4 nm). The filaments are arranged regularly, but no Z-discs or similar structures were seen. Bairati suggested two functions for the pigment layer; protection from radiation and osmoregulation. The function of the myoid layer he considered to be tonic, maintaining intratesticular pressure and furthering the progression of stages toward the exit, but not peristaltic. The testis appeared immobile. Although this type of sheath may, of course, be present only in some species, the finding of a myoid element relates it to the sheaths in many other phyla and is particularly reminiscent of mammals. Bairati defined the wall of the insect testis as 'a membrane containing and enveloping the germinal tissue *without* entering into a stable association with it'.

Whatever the function of the testicular envelope, it appears to be maintained when the testis is kept *in vitro*. Since Goldschmidt's experiments, whole insect testes have been kept in tissue culture repeatedly and it has been shown that cellular differentiation within them may parallel that *in vivo* (Fowler, 1973).

Germ cells

In most insects spermatogonia and spermatocytes develop in the pupal and nymphal stages, and the testis of the imago contains only spermatids and spermatozoa, or in species with a very short adult life (e.g. mayflies and caddis flies), only spermatozoa. On the other hand, there are species with a relatively long adult life (e.g. many bettles) in which meiotic and even premeiotic stages are found in the adult testis.

It has been known for many years that the germ cells of insects develop in clones. In a majority of species the derivatives of a single spermatogonium remain together in a well demarcated cyst, and this fact has provided unequivocal evidence for the clones which were described in clear detail in many classical papers, e.g. by Davis (1908) and Mohr (1914) in Orthoptera, Henderson (1907) and Hegner (1914) in Coleoptera, and Lutmann (1910) in Trichoptera. It was not so evident, although occasionally seen (e.g. Hirschler, 1953) before the advent of electron microscopy that the encysted germ cells are connected by intercellular bridges. Probably the first ultrastructural demonstration of bridges, "fusoms", between spermatogonia, spermatocytes and spermatids was produced for *Drosophila* (Meyer, 1961). Bairati (1967) in another electron microscopic study of that species emphasized 'the syncytial state of the germ cells', and pointed out that the intercellular connections are limited in early stages because the bridges are narrow, but that they become much wider in spermatids. During later stages of spermiogenesis an "individualization process" takes place in which the syncytial state is dissolved (Tokuyasu, Peacock & Hardy, 1972). The process eliminates not only intercellular bridges but also removes redundant organelles, excess membranes and cytoplasm in a separate lobe which is cast off from each spermatid.

The bridges presumably develop first in secondary spermatogonia as the result of incomplete cell divisions. After three or four divisions the interconnected spermatogonial clone typically assumes the configuration of a "rosette" (Henderson, 1907; Knaben, 1931, etc.) and in many cases the scale-like clones become stacked in series in the apical end of the follicle (Plate 10*b*). The cysts become more spherical in the spermatocytic stage, and finally elongate during spermiogenesis when the flagella of the spermatids develop. The arrangement of the cells and the orientation of the bridges of the 64 spermatocytes in the cysts of a lepidopteran have been explored in a unique investigation by Hirschler (1953). He came to the important conclusion that the spermatocytes in a clone while essentially genetically alike, are not identical because of uneven distribution of spindle-remains, centriolar apparatus and possible other cytoplasmic components.

The huge variety of patterns in insect spermiogenesis has been barely glimpsed in a great number of electron microscopic investigations, which have been partially reviewed by Phillips (1970). In describing the stages of spermatogenesis of *Locusta migratoria* from an ultrastructural point of view, Szöllösi (1975) has developed a useful list of ten stages which can easily be modified for other species:

Stage 1. Mitochondria cluster at one pole of cell; budding of flagellum.

Stage 2. End-to-end fusion of mitochondria forms 'Nebenkern'.

Stage 3. Centriolar adjunct arises and proacrosomal granule is formed.

Stage 4. Nebenkern divides into two derivatives.

Stage 5. Chromatin condenses into fibers and a perinuclear 'chromatoid body' appears. Accessory tubules form in the axoneme of the flagellum.

Stage 6. Chromatin fibers form lamellar sheets, and acrosome is completed; cytoplasm begins to slough off.

Stage 7. Chromatin sheets show honeycomb pattern; core appears in accessory and central tubules of axoneme.

Stage 8. Spermatid bundle is formed; intercellular bridges disappear; concentric rows of microtubules are aligned around nucleus and mitochondrial derivatives.

Stage 9. Complete condensation of nucleus; formation of spermatodesm (see Glossary).

Stage 10. All microtubules disappear.

This list may serve as an outline of the features of typical spermiogenesis in an insect.

Interesting deviants of the meiotic process occur in many insects, but in particular among the dipteran suborder Nematocera. Part of these have been excellently reviewed by Metz (1938). In most of the aberrant forms certain features are similar: no bivalents are formed, the first meiotic division is unequal, and only one or two spermatids are produced by each primary spermatocyte. This is often associated with the production of unisexual broods, the females of which are exclusively either male- or female-producing. An instructive example is *Miastor* (White, 1946). In this insect four spermatogonia are contained in a cyst before the last premeiotic division. Each nucleus has 48 chromosomes, the octoploid number. By the end of larval life each testis contains 256 spermatocytes I in 32 cysts with 8 cells each. In the first meiotic division the classical stages are not recognizable. Each octoploid spermatocyte gives rise to 2 cells, one containing 6 and another 42 chromosomal units. Only the first of these divides again and gives rise to two haploid spermatids; the other degenerates without division. In a related species, *Taxomia taxi*, decaploid primary spermatocytes similarly produce two haploid spermatids (White, 1947).

The control of germ cell multiplication in insects has been explored in some detail in *Drosophila melanogaster.* The process of spermatogenesis proceeds in larval abdomens isolated by ligature in the same way as in controls which undergo metamorphosis (Hadorn *et al.*, 1964) even though, as has been mentioned, the somatic cells of the testicular envelope do not differentiate under these circumstances. On the other hand, when larval testes were transplanted into adults and host glands were either activated or extra glands implanted, it could be demonstrated that Median Neurosecretory Brain Cells (MNBC), and not the corpus allatum, are stimulating germ cell multiplication (Garcia-Bellido, 1964*a*). This control is probably indirect. The MNBC act on protein assimilation which in turn maintains the autonomous protein synthesis in germ cells without which cellular multiplication does not take place.

When spermiogenesis begins in the larval testis, spermatogonial divisions are inhibited (Garcia-Bellido, 1964*b*), although the differentiation of older cysts continues. It is conceivable that there is competition over a common necessary substrate between proliferating spermatogonia and differentiating spermatids, and indeed the performance of developing germ cells in isolated abdomens suggests that maturing meiotic and postmeiotic cells can survive with much less substrate than spermatogonia. However, a specific inhibiting effect of spermatids on the formation of new cysts has not been excluded. In what is somewhat of a key paper Dumser & Davey (1975) described that the spermatogonia of *Rhodnius* exhibit a basal level of division rate in the absence of morphogenetic hormones. Ecdysone either naturally induced by feeding or injected, approximately doubled the mitotic index as measured by metaphase accumulation after administration of colchicine. Juvenile hormone production in molting larvae or application of a substance mimicking the effect of the hormone abolished this ecdysone effect but did not affect the basal level of the mitotic rate. (The natural induction of ecdysone had no effect on the duration of the meiotic prophase.) The ecdysone effects described have been implied in the work of several authors cited.

Cysts and cyst cells

As a general rule, the greatest part of sperm development in insects occurs in "spermatocysts" in which germ cells develop more or less synchronously surrounded by a capsule of somatic cells. In Diptera such cysts are not conspicuous and may occur in a modified form or not at all, which will be discussed below. In Protura they may not be present (Berlese, 1910), which is of interest in view of the fact that the systematic position of this group is at present still a matter of opinion (Imms, 1957).

Cysts begin to be formed during the spermatogonial stage and by the time spermatocytes have differentiated a lumen appears. Typically a fully grown cyst is a hollow sphere of 32–64 germ cells, surrounded by an envelope consisting of one or more "cyst cells". There is general agreement that each cyst contains one clone of germ cells derived from a single primary spermatogonium (Snodgrass,

1935). This suggests two possible modes of origin. Either a single primary spermatogonium becomes surrounded and isolated by somatic cells, or secondary spermatogonia remain closely associated and are enveloped as a group after one or more cell divisions. Both conditions apparently occur in nature. Holmgren (1901, on Coleoptera), Davis (1908, on Orthoptera) and Zweiger (1906, on Dermoptera), for instance, have described and illustrated single primary spermatogonia surrounded by a single cyst cell. On the other hand, Ammann (1954, on Lepidoptera) observed that cysts formed around groups of 8 or 16 interconnected spermatogonia. Younger germ cells were unencysted.

Detailed descriptions of cyst formation have been given by Davis (1908), Demandt (1912) and Nelsen (1931). According to Davis, outside the corona of spermatogonia surrounding an apical cell, a number of somatic cells are found which resemble 'connective tissue cells'. When a primary spermatogonium divides, one of the daughter cells is usually forced out of the corona. This secondary spermatogonium becomes intimately associated with one or more somatic cells which form an investment around it. The association persists until the spermatozoal stage. Demandt found in *Dytiscus* that the apical region of the testis is filled with cells varying widely in nuclear size and chromatinic pattern (Plate 9a). The conclusion seems justified that they represent a mixture of germ cells in various stages of development and of somatic cells. Cell borders are indistinct. In the adjacent zone all cells have definite border lines but somatic cells are still not identifiable. About 1½ mm from the apex the spermatogonia appear arranged in rosettes each of which is enveloped by two cyst cells. The spermatogonia continue to multiply within the cysts. At a stage when they were identified as spermatocytes by Demandt, the cells lose their cohesion, no longer form rosettes, and the cysts become very irregular in outline. The cyst cell nuclei become much larger than the nuclei of the germ cells. Scattered degenerating spermatogonia are frequent and in the zone of spermatocytes whole cysts are often necrotic. It appears that the degeneration products are taken up by cyst cells which grow rapidly and tend to fill the spaces created by the dissolution of cysts. Demandt was under the impression that degeneration was the more frequent the less well nourished the animal. Gradually, the whole degenerating mass is pushed toward the center of the testis where it forms a strand of debris, or as Demandt called it, 'Nährsubstanz'. The flagella of developing spermatids first point toward the hollow center of the cyst, later toward the center of the testis (Plate 9b and c). Although the cyst wall regresses, the orientation of spermatids remains unaltered at first. Later they become reoriented parallel to the axis of the testis and move head first into the epididymis. Remains of cysts form a large yellow plug which fills almost the whole cavity in the winter testis of *Dytiscus*. In Orthoptera, Davis (1908) and Nelsen (1931) did not observe degenerative phenomena, but Baumgartner (1929) showed that the distal end of the testis of the cricket contains a 'nutritional mass' of debris, similar to that seen in *Dytiscus*.

Although cysts always arise before the growth period of the spermatocytes, the cyst cells do not reach the highest degree of differentiation until spermiogenesis. Gilson (1884) was able to study the nucleus of cyst cells only in the growth stage of germ cells 'because it is only at the end of this stage that this problematic body becomes visible'. Usually a very large nucleus was found at the head of each bundle of spermatids. In Coleoptera two nuclei are placed regularly at the sides of each bundle. Some Hemiptera have one, some several, nuclei for each bundle. In Orthoptera, Gilson did not succeed in finding cyst cell nuclei, but he suspected that they might be hidden among the very voluminous bundles of spermatids.

In an important cytological and cytochemical study of the cyst wall Anderson (1950) investigated spermatogenesis in the Japanese beetle, *Popilia japonica*. During spermiogenesis the cyst cells are large and flat, and form an envelope which closely follows the outlines of spermatid bundles. In the phase of nuclear condensation the sperm heads turn toward the walls of the cyst and push deeply into the cyst cells. Anderson thought that the acrosome actually penetrates into the lining cell, but this is most improbable in view of massive electron microscopic evidence to the contrary in many other insect species. The cyst cells show strong basophilia due to RNA. They are also rich in glycogen which at later stages is concentrated around the heads of the invaginating spermatids. The alkaline phosphatase reaction (Gomori) shows positive granules only at the tip of the spermatid and in the immediately adjacent regions of the cyst cells. These appear to phagocytize remains sloughed off from the spermatids.

Anderson observed living spermatids and saw that they assumed a 'regimented appearance' due to a coordinated undulation of their tails. The cyst was somehow propelled through the crowded follicle into the funnel of the vas efferens. The spermatozoa were released from the cysts after they entered the vas and the abandoned cyst cells began to degenerate there. Basophilia and lipids disappeared from their cytoplasm, but glycogen increased and persisted until the cells dissolved in the female tract after copulation. Anderson suggested that the glycogen serves as nutrient for the spermatozoa in the seminal receptacle. In another beetle, *Prionoplus reticularis*, Edwards (1961) found that the cyst cells did not leave the testis at all, but broke down and dissolved while still in the testicular cavity; some formed a group of large cells with polyploid nuclei which persisted until senescence.

Phagocytosis by cyst cells appears to be widespread. It was described for Coleoptera by Henderson (1907), Wiemann (1910a) and others; for Orthoptera (see above); and less frequently for Lepidoptera (Goldschmidt, 1917; Ammann, 1954). In Hemiptera great enlargement and polymorphism of the nuclei has been observed regularly (Tannreuther, 1907; Bowen, 1922a; Poisson, 1936). Bowen claimed that the cyst cells removed the cytoplasmic remnants of spermiogenesis so completely that none ever got into the vas deferens.

Hughes-Schrader (1946) has described an additional aspect of cyst cell behavior in an iceryine coccid. In this animal, the tails of the spermatids come to lie in an emerging lobe of the cyst. The germ cells are arranged in bundles which are initially subdivided by processes of the cyst cells. These cytoplasmic septa finally rupture and at the same time the wall of the cyst stiffens as demonstrated mechanically. All spermatids in a cyst are brought into rigid alignment. This behavior of the envelope has not been described in other insects, but reminds one of arachnids (see Chapter 6.1). However, cellular processes which pervade and subdivide the interior of a cyst appear to be very prominent in Hymenoptera where they were first described in detail by Meves (1907) for the honeybee. In this species the cyst cells, which are continuous with the inner lining of the testis, are extremely rich in mitochondria. As the cysts grow, the cyst cells multiply and send prominent processes filled with mitochondria into the cyst, enveloping individual or small groups of spermatocytes. Meves' drawings (Fig. 4, p. 10) appear to be good evidence, but electron microscopic accounts have not confirmed them as yet. Hoage & Kessel (1968) and MacKinnon & Basrur (1970) do not mention nor depict any characteristic processes of cyst cells. On the other hand, Smith (1968) shows electron microscopic pictures of lepidopteran cysts demonstrating processes of 'follicle (or 'sustentacular') cells which meander between the spermatids, and perhaps play a part in regulating development in addition to, or instead of, their generally assumed nutritive role'. An extension of the surface of the cyst cells associated with an increase of contact with the germ cells appears to be the highest degree of differentiation of the cyst arrangement in insects, but at present there is insufficient information to indicate how widely distributed it is.

The environment of the cysts varies considerably in different phases of development and in different species. In early stages cysts are usually tightly packed. Later they may seem to float in a fluid-filled cavity (Zick, 1911). However, they may also be suspended in a loose, interstitial network of "trophocytes" continuous with the inner lining of the follicle as in the milkweed bug (Heteroptera; Bonhag & Wick, 1953). In still other cases the cysts remain densely crowded throughout spermatogenesis and no obvious interstitium develops (Meves, 1907; Tannreuther, 1907).

In Diptera, cysts are not obvious. Cholodkovsky (1905) recognized in *Leptis* that the interior of the follicle contained some interstitial cells derived from the epithelial lining of the inner tunica. These appeared to partition the germ cells into groups. In *Dolichopus* and *Volucella* he described 'yolk containing' vacuolated cells, which he regarded as nurse cells, distributed among spermatocytes and spermatids. In *Laphira*, he observed cyst-like structures but did not mention any investment. He did, however, describe branching septa between these 'cysts' which were connected with the inner lining of the follicle. Between the spermatogonial and the spermatocytic region which he called the 'cyst zone', he found a vacuolated and granular mass staining with carmine. He speculated that this might be a 'conglomerate of fused nurse cells'. The observation was confirmed by Kenchenius (1913). Lomen (1914) mentioned no 'cysts' in *Culex pipiens*. He did notice large cells among cellular debris in the spermatozoal zone and suggested that these might be derived from germ cells. In *Aedes aegypti* Jones (1967) described 'compartments', maximally 24 per testis, which he called 'spermatocysts'. Each of these contained 500 or more synchronous spermatids.

Drosophila has been investigated particularly thoroughly by a number of investigators (for review see Cooper, 1950). Aboim (1945) came to the conclusion that 'interstitial cells' somewhat tenuously delimit groups of germ cells which in other insects are enclosed in definite cysts. These cells are connected with the tunica propria of the testis, as in many other dipteran species. However, Aboim's experiments with agametic testes showed that interstitial cells are not derived from this layer but from a mesodermal group of cells, the "apical cells" which will be described below. Although the resolution of the light microscope did not permit detailed observations on the distribution of the fine investing processes of interstitial cells, Aboim expressed the conviction that they delimit germ cell territories. This appears to have been conclusively confirmed by electron microscopic investigations. Bairati (1967) has described the ultrastructure of interstitial cells in detail. They are easily recognized in electron micrographs by their volume, characteristic shape and richness in organelles. They are less electron-dense than the germ cells and in this respect resemble the Sertoli cells in mammals. Bairati described the development from the polygonal cells in the apical part of the testis to the thin, flattened cells ('cellules cistiches') enveloping bundles of spermatids. In the terminal zone degenerative cysts appear. Here the cyst cells contain many organelles and large inclusion bodies and finally lose continuity, i.e. the cysts dissolve.

A different cell type has been described in *Drosophila* as 'nutritive cells' (Aboim, 1945; Cooper, 1950). These are giant cells scattered along the wall of the distal part of the follicles. Bundles of advanced spermatids become implanted in them. Fujimura, Masuda & Ueyama (1957) showed electron microscopically that each 'nutritive cell' is well defined by cell borders and contains much rough endoplasmic reticulum (ER) and great numbers of multivesicular bodies which today would be interpreted almost certainly as lysosomes. It now appears that these cells are derivatives and highly differentiated stages of interstitial or cyst cells (Bairati, 1967).

An interesting correlation between germ cells and "nurse cells" was discovered by Montgomery (1910) in the genus *Euschistus* of the family Pentatomidae (Hemiptera). The testes of these animals consist of six lobes or follicles arranged in a series side by side. The lobes may be identified by number and number '1' assigned to the lobe on the side toward which the efferent duct opens (Bowen, 1922*b*). Montgomery observed that all spermatocytes in lobes 4 and 6 are several times larger by volume than those in the other lobes. The large spermatocytes develop into large spermatozoa. Spermatogonia are of equal size throughout and size differences do not become measurable until the growth period of the primary spermatocyte begins. Germ cells are enveloped by cyst cells and between the cysts are thin septa formed by 'nurse cells' resting upon the inner lining of the lobe. In lobes 4 and 6 cyst and nurse cells are twice as large as in the other lobes.

Montgomery concluded that the larger cells contribute more nutriment to the developing spermatocytes which grow larger in consequence, but Bowen (1922*b*) pointed out correctly that this can hardly serve as an explanation for the dimegaly, because the cause of the unequal growth of nurse cells remains unexplained. Montgomery himself did not succeed in showing any obvious differences in blood supply or tracheal arborizations between the lobes. Bowen established that the phenomenon is less unique than Montgomery had thought. In the family Pentatomidae polymegaly of spermatozoa and nurse cells is frequent. Usually the large cells occur in lobes 3 and 5, or 4 and 6. Sometimes unusually small spermatocytes occur in another lobe. Although the nurse cells are not necessarily responsible for the polymegaly, and may, in fact, be large because they are stimulated by an unusually high demand of the spermatocytes, the correlation itself is not in doubt. It indicates an intimate metabolic relationship between cyst cells, nurse cells and germ cells. The cytology of the heteroploid germ cells has been discussed most recently by Schrader (1945*a*, *b*).

Apical cells

A remarkable and, perhaps, unique feature of insect spermatogenesis is the association of primary spermatogonia with a large cell or cellular complex at the apex of the testis or of each follicle. This 'apical cell' (Grünberg, 1903) was first described in Lepidoptera by Spichardt (1886) who named it 'Keimstelle' because he considered it the source of germ cell production. Verson (1889) found that in each follicle of *Bombyx* 'there is a single, large germ cell from which gradually all organized structures of the follicle take their origin'. This sweeping view in combination with a dramatic, oversimplified woodcut in which the apical cell appeared in black (Fig. 19), aroused great interest, and Verson's name has ever since remained associated with this cell, although the inappropriateness of the eponym was pointed out already by Grünberg (1903) who gave a much more adequate description of the structure. According to him the apical cell of Lepidoptera develops in late embryonal stages in both sexes but soon becomes larger in the male. Presumably it is derived from primordial germ cells, but does not give rise to spermatogonia. Primary spermatogonia surround the cell and extend processes toward it which make contact with its surface (Plate 10*a*) and interdigitate with processes from the apical cell. As the follicle develops, some of the primary spermatogonia degenerate and are taken in by the apical cell. Initially the cell rests against the inner lining of the follicle. Later the tunica propria invaginates and the apical cell moves into the center of the follicle on a stem formed by the invaginated lining. The follicular lining is very tenuous at the point of contact with the apical cell and Grünberg observed fine striations from the cytoplasm of the cell into the attenuated part of the membrane. As development proceeds, the apical cell grows, its cytoplasm becomes more granular and its nucleus polymorphic. Degenerating spermatogonia sometimes appear to lie in bays of the nucleus. Grünberg compared

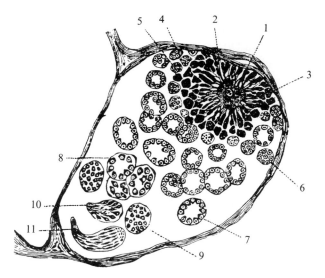

Fig. 19. Schematic representation of a follicle of *Bombyx mori*, in which the apical cell (1) was first presented. 2, Smaller surrounding cells (spermatogonia); 3–11 successive stages of developing cysts. (After E. Verson (1889). *Zool. Anz.* **12**, 100–3.)

the apical cell with the follicular cells of the ovary, which also extend processes toward the gametes as, for instance, in *Dytiscus*. In almost mature caterpillars the apical cells lose their connections with the tunica propria and become vacuolated. In the imago the nucleus fragments and the cell dies.

There are considerable species differences even within the Lepidoptera. Only in *Bombyx* is the stem formed by the invaginated epithelium very conspicuous. Only in *Gastropacha* did Grünberg observe very long processes which reached beyond the adjacent spermatogonia far out among the spermatocysts. He became convinced that the morphology of the apical cell indicates its role in supplying the spermatogonia with nourishment and mediating between them and the follicular wall. In addition, the cell phagocytizes degenerative germ cells. It functions only during the larval period. In the female it does not enter into any obvious association with the germ cells, remains undifferentiated and degenerates already in the larva. Most of Grünberg's findings have been confirmed by Zick (1911) who developed the concept that the apical cell may control the timing of spermatogenesis and suggested that spermatogonia are able to begin development only after they leave the immediate association with the apical cell. Buder (1917), who again described apical and germ cells in great detail, expanded Zick's theory and postulated an inhibiting factor emanating from the apical cell. Other descriptions of confirmatory nature have been published by Schneider (1917), Ammann (1954) and Wittig (1960). Ammann stated that the growth of the apical cell in the very young caterpillar is the first sign of sexual differentiation of the male.

Muckenthaler (1964), in one of the very few studies in which modern methods were applied to the exploration of the apical cell, found in the grasshopper that [³H]uridine and [³H]adenine were incorporated vigorously into the nucleus but only minimally into the cytoplasm. He concluded that there is a high rate of RNA synthesis in the apical cell at all times during spermatogenic activity.

Apical cells occur in many orders of the Insecta. The Lepidoptera represent the classical case and it appears that all species investigated have revealed them. Only in *Tischeria* was the existence of apical cells denied by Knaben (1931). However, he studied only the last instar of the larva and there is the distinct possibility that the apical cells in this species degenerate somewhat earlier than usual.

In other orders it is not possible to obtain a fair picture of the occurrence of apical cells. The number of cases reported is small, and pertinent information was discovered only for 11 out of 29 orders. The most complete documentation is available for Orthoptera (Davis, 1908; Schellenberg, 1913; Nelsen, 1931; McClung, 1938; and others). Characteristic apical cells were found in all species investigated. In Hemiptera their existence was convincingly demonstrated by Kornhauser (1914), Bonhag & Wick (1953) and Masner (1965). Kornhauser observed that the apical cell may occasionally be involved in the phagocytosis of spermatocysts. In *Euchenopa* cysts frequently degenerate and apical cells can be seen to extend large processes around them. Bonhag & Wick described what they called an 'apical complex' in the milkweed bug, *Oncopeltus*. The structure contains 9–11 nuclei in a large cytoplasmic territory. Cellular processes radiate toward the surrounding primary spermatogonia. Some of these are in a state of partial dissolution 'suggesting that their breakdown may be due to the activities of the apical complex'. Masner, like Kornhauser, found mononucleate apical cells (in *Adelphocoris*). He considered all germ cells in contact with apical cells as primary spermatogonia.

In Protura, Berlese (1910) demonstrated a fairly typical apical cell in the recurved end of the single tube of the testis. Illustrations of Perrot (1934) indicate that apical cells may exist in Phthiraptera, although they have not been described.

Diptera usually possess apical cells, but apparently these differ widely in the degree of differentiation. Cholodkovsky (1905) found that they were best developed in *Laphira* and *Asilus*, which was confirmed by Metz & Nonidez (1921), and least developed in *Musca domestica*, but in this species they may have been more clearly differentiated at earlier stages. He saw evidence for secretory and phagocytic function. Kenchenius (1913) observed only with low magnifications, and with these he discovered apical cells in Syphidae, but not in Muscidae and Leptidae. His statement that apical cells have disappeared in the imago of most Diptera disagrees with the conclusions of Cholodkovsky and others. In *Culex*, Lomen (1914) found no apical cells, but a well demarcated mass of tightly packed primary spermatogonia sunk into the capsule at the apical end of the testis. In *Psychoda*, Friele (1930) found apical cells during the larval period. She rejected the concept that these cells are nurse cells, because their size is small when compared to the spermatogonia. Aboim (1945) observed in *Drosophila* at the proximal end of the testicle a group of a few small cells, sometimes well demarcated, sometimes of uncertain outline, which he called 'apical cells'. He rendered animals agametic with ultraviolet irradiation and produced evidence that the apical cells develop independently of germ cells and give rise to interstitial cells. However, the cells he described differ greatly from the characteristic apical cells of Lepidoptera and Orthoptera. They appear to be similar to those which Lomen observed in the mosquito and called 'spermatogonia'. Cooper (1950) considered the evidence in *Drosophila* sufficient to make the statement that 'the apical cell, the interstitial cell (hence the septal cell), and the nutritive cell may represent three steps in the evolution of the same cell line, all perhaps devoted in a trophic fashion to the welfare of the developing germ cells'. Bairati (1967) in his electron microscopic study appears to have confirmed this view. He considered the cells mingled with primary spermatogonia in the apical end of the testis of *Drosophila* as interstitial cells in an early stage of development and had no doubt that they were indeed the 'apical cells' of Aboim.

In Coleoptera, typical apical cells were described by Holmgren (1901, 1902) for *Staphylinus* and *Silpha*. In the last he found one to three cells per follicle. In other Coleoptera typical apical cells are absent, but the apex of the follicle may contain special structures the nature of which is not clear. Demakidoff (1902) described in *Tenebrio* an ellipsoid body of somewhat fibrous appearance and with three to five nuclei, in the location of an apical cell. He called this structure a 'lens' and it has later often been referred to as "Demakidoff's lens". He concluded that the lens is not homologous to the apical cell. Demandt's (1912) illustration of the apical region of the testis of *Dytiscus* does not reveal any apical cells (Plate 9a). In *Leptinotarsa signaticollis*, according to Wiemann (1910b), there is a terminal cap of epithelial cells at the apical end of the testis. The cap is sharply delineated from the primary spermatogonia which abut against it. At the center of its free surface a plug of degenerative material slowly forms until it almost fills the cavity of the follicle. Wiemann believed that this plug develops from the cells of the cap and serves to nourish spermatozoa at a late stage of maturation after they have been released from the cysts. Hegner (1914) confirmed the existence of a terminal cap in *Leptinotarsa decemlineata*, the Colorado beetle, but made no mention of the degenerative mass. There are points of similarity in the descriptions of Demakidoff's lens and the terminal cap, and one wonders whether their products, the fibrous strand and the plug of cellular debris, may not also be related. The subject should be tempting to investigators with an electron microscope.

Naisse (1970) has suggested that the apical tissue in *Lampyris noctiluca*, a beetle, produces an androgenic hormone which under the control of a neuroendocrine complex stimulates the development of a male gonad. If true, this function of apical cells cannot be general because it is not at all certain that differentiation of sex in insects is usually mediated by hormones (Laugé, 1970).

Menon (1969) studied two tenebrionid beetles with light and electron microscope. One species, *Tenebrio molitor*, is the same as Demakidoff (1902) investigated. Menon, in contrast to Demakidoff, demonstrated that there is a characteristic 'apical' tissue in these beetles which is ultrastructurally easily distinguished from neighboring germ cells by small nuclei, highly convoluted cytoplasmic membrane, dense cytoplasm, profuse granular endopasmic reticulum, large aggregates of glycogen granules, numerous small mitochondria, well developed Golgi centers and remarkable arrays of microtubules. Menon observed that many of these features resemble those of the androgenic glands in crustaceans, but not those of androgenic cells in vertebrates. While endocrine function is not indicated, there clearly appears to be some secretory activity. The function of the microtubules remains uncertain. No phagocytosis was observed in this case.

In some pertinent descriptions on Hymenoptera (Meves, 1907), Dermaptera (Zweiger, 1906), Trichoptera (Lutmann, 1910) and Thysanura (Charlton, 1921) there is no evidence of apical cells or apical complexes.

Goldschmidt (1931) in a paper which contains some excellent photographs of insect testes was able to show what happens in certain intersex moths when an ovary transforms into a testis. Spermatogenesis begins, preceded by degeneration of oocytes and egg follicles, at the periphery of the gonad. While normal male moths possess typical apical cells, the intersex animals converting into males show no apical cells in connection with spermatogonia. In fact, the undifferentiated female apical cells have no relationship to the foci of male development. In male intersex animals converting into females, the apical cells surrounded by a corona of spermatogonia are among the last testicular structures to be affected by the transformation. From these results, Goldschmidt drew no conclusions about the significance of the apical cell, but Carson (1945) suggested that they were a clear demonstration of the fact that the apical cell is unimportant in the development of spermatogonia or any other phase of spermatogenesis. His argument, however, is not conclusive. After careful scrutiny of the results as described by Goldschmidt, it becomes clear that secondary male development takes place in a very complex environment in which degenerative products of female structures in regression play an important part among other factors. Something in this environment may serve as a substitute for whatever the apical cell normally supplies.

In a comparative study of the apical cell, Carson brought forward what he considered convincing evidence that mitochondria are transferred from apical cell to spermatogonia in the robber fly, *Erax estuans*. In this species all primary spermatogonia can be seen to have a long, slender process which extends 'toward and into' the apical cell. Where the process touches there is 'almost invariably a mass of heavily-staining granular mitochondria some of which material in many cases can be seen to have entered the process'. Carson thought of the apical cell as nutritive, although of secondary importance. In interpreting Goldschmidt's results he finally concluded that the apical cell has no 'fundamental significance' because it occurs irregularly and its secretions are dispensable. This strong opinion was in part directed against the opposite claim of McClung (1938) which is of historical interest. McClung was convinced that he had discovered in the apical cell (of the grasshopper) the means by which 'the experience of the individual, reflected in peculiar materials deposited in the apical cell, is brought into effective relation with the racial organisation represented by the chromosomes of the germ cells'. Obviously, if this grandiose concept has even a grain of truth in it, there must be similar mechanism in most animals, and McClung did not hesitate to assert that 'doubtless careful studies will show definite and suggestive relations between somatic and germ cells'. He categorically declared that there is nothing to suggest that the apical cell is an ordinary nurse cell.

A few remarks must be added about the origin of the apical cells, although this question also cannot be answered definitely at present. Many of the earlier investigators tended to the belief that the apical cells were derived from 'primordial' germ cells through unequal cell divisions (see Grünberg, 1903). Recent conclusions have been much more cautious (Ammann, 1954). In the case of the typical apical cell which occurs in Lepidoptera and Orthoptera there simply is no clearcut evidence available for their early embryonal derivation. In Diptera, in particular in *Drosophila*, there has been some confusion because of the small size of the apical cells and their superficial similarity with spermatogonia. But Hannah-Alava (1965) has shown that their nuclear morphology is distinctly different, and Aboim (1945), as explained above, demonstrated experimentally that in *Drosophila* apical cells develop normally in animals in which the germ cells have been wiped out.

An adequate summary statement on the significance of apical cells in the insect testis cannot be given at present. As they always occur in direct association with the early stages of germ cells, the spermatogonia, their role must be thought to lie first of all in this association. The similarity to the relationship between follicular cells and eggs is obvious, and some answers will probably be found in applying experimental methods capable of clarifying the possible transport of metabolites between apical cells and spermatogonia. The present state of our knowledge may be summed up as follows:

(1) The apical cell in its characteristic form is a large, mononucleate cell occurring singly in each testicular follicle and found regularly in Lepidoptera, Orthoptera and Hemiptera. Other apical structures, sometimes similar, sometimes very dissimilar, have been described for many other insects, but their homologies are uncertain.

(2) The apical cell develops in embryonal life and is the earliest sign of male differentiation of the gonad. It is surrounded by primary spermatogonia which usually are in intimate contact with it. It may play an indirect role in the spermatogonial renewal system (see Chapter 9). The apical cell degenerates when spermatogenesis ceases; in Lepidoptera in prepupal stages, in Orthoptera and Diptera usually in the imago.

(3) There is direct evidence that the apical cell may phagocytize the remains of degenerating spermatogonia, and suggestive evidence that it supplies spermatogonia with nutritive substances. It does not give rise to germ cells, but in Diptera, at least, it may give rise to interstitial and cyst cells.

Various invertebrate phyla

Information on many invertebrate phyla is too scanty to justify separate chapters. They will be reviewed here collectively despite their great diversity. No detailed descriptions appear to exist on Acantocephala, Ectoprocta, Phoronida, Brachiopoda, and Sipuncula, but some fragmentary data will be mentioned.

Mesozoa

Only the aberrant group of the Dicyemida will be discussed here, which is probably better designated as a separate phylum (Kozloff, 1969). The animals have a remarkable and unique mode of gametogenesis which takes place in a large axial cell of the sexually reproducing form (Whitman, 1883). Female and male gametes develop within the axial cytoplasm. The spermatogenic cells lie in groups called 'infusorigens'. The process was first described superficially by Nouvel (1947) and later investigated in some quantitative detail by Austin (1964). In *Dicyema aegira* no more than five male germ cells, one spermatogonium and four spermatids, are ever observed in one infusorigen. Presumably this indicates that a spermatogonium does not divide until the spermatids of the previous generation have left the cell group as spermatozoa in order to fertilize a mature oocyte in the immediate vicinity. After spermiation the spermatogonium probably divides quickly because single spermatogonia are rarely seen. It apparently divides only once in an unequal division giving rise to one primary spermatocyte and one spermatogonium. Perhaps, this represents a "spermatogenic cycle" with the smallest possible cell number. Ridley (1969) investigated infusorigens with the electron microscope. He was unable to find any cytological differences between spermatogonia and spermatocytes, but as he did not describe stages of single cells, it is not certain that he dealt with any spermatogonia at all. The cells which he observed had relatively large nuclei, a distinct nucleolus and peripherally located chromatin. No meiotic phenomena were seen. The spermatid nuclei are smaller (4 μm) and have dense cytoplasm. The tailless spermatozoa show a granular central mass of chromatin which is apparently not bounded by a membrane, but a plasma membrane was clearly identified by Ridley, in contrast to observations by Austin.

Platyhelminthes

Although the number of species in this phylum is great, very few have been studied with regard to spermatogenesis. The great majority of these come from two of the three classes (classification of Dawes; Rothschild, 1965); the tapeworms (Cestoda) and the flukes (Trematoda). Hyman (1951) has briefly reviewed the gonadal structures of the phylum which in Turbellarians alone range from ill defined (e.g. order Acoela) or well delineated (e.g. order Polycladida) multiples to single pairs (e.g. order Rhabdocoela) of organs. Successive steps of organization of gonadal tissues can be traced in this phylum. The formation of a tunica around the testis is probably incomplete in many species, e.g. in planarians (Schleip, 1907), but well organized and containing myoid elements in others, e.g. trematodes (Child, 1907). Lindner described the 'testicular vesicles' ('Hodenbläschen') in one trematode as cavities in the parenchyma, without a proper wall. The multiple compartments or cellular aggregations of male germ cells in flatworms frequently are called testes, but also testicles or follicles. In size they are usually, e.g. in planarians, comparable to lobules or seminiferous tubules in molluscs or mammals, and in these cases do not deserve the name of "testes". But in certain platyhelminthes, particularly in trematodes, the follicles fuse into true testes which may be organized in lobules and lobes.

Gametogenesis in the class Turbellaria has been reviewed by Henley (1974) who emphasizes the great diversity of structures connected with the complex genital apparatus. The spermatozoa may possess undulating membranes, bristles, paired flagella etc., and vary in length from less than 50 μm to 400 μm. According to Lentati (1970) the fertilizing spermatozoa are usually haploid although polyploid biotypes exist in which, as for instance in *Dugesia benazzii*, excess sets of chromosomes may gradually or in total be eliminated during spermatogonial or meiotic divisions.

A classical description of a spermatogenic follicle was given by Schleip (1907) for *Planaria gonocephala*. This animal possesses hundreds of follicles which become only gradually delimited by a membrane as spermatogenesis advances. The follicles grow as they mature, and several stages appear in them, the older ones in the center. Schleip described clusters of approximately 32 spermatids connected by a large spherical mass of cytoplasm, a 'cytophore'. He considered spermatogonia as syncytial but this was doubted by Rappeport (1917) who otherwise confirmed most of Schleip's findings for *Planaria alpina*. Most investigators have described cytophores in turbellarians with the exception of Weygandt (1907) who found the cell divisions in *Plagiostomum girardi* complete and the gametes separate. However, he also described the spermatids in clusters associated with masses of residual cytoplasm.

Frauquinnet & Lender (1973) have shown that the ultrastructure of the spermatogonium in planarians is identical to that of neoblasts, the pluripotent cells in the parenchyma. It has long been believed that the germ cells in planarians are derived from these cells. This was elegantly demonstrated by Ghiradelli (1965). He amputated the cephalic region of mature animals which causes the gonads to regress. As the head regenerates the gonads reappear in their previous locations. The formation of testes is always initiated by an accumulation of neoblasts. Fedecka-Bruner (1965) has shown the same sequence after the testes were destroyed with X-rays.

For Trematoda there exist a few thorough descriptions of testes and spermatogenesis and a relatively large number of papers containing fragmentary data. According to Dingler (1910) the testes of *Dicrocoelium lanceolatum* have a peripheral layer of spermatogonia among which are occasional large cells of unknown nature. Each spermatogonium divides three times and the resulting cells remain connected by single processes which meet in a center, but no large cytophoric mass is seen. Spermatocytes I occur in connected groups of 8, spermatocytes II in groups of 16 and spermatids in groups of 32. At the end of spermioteleosis the spermatids are released and a residual body remains with 32 cytoplasmic masses adhering in a "morula" which finally degenerates. Lindner (1914) found that germ cells in the testes of *Schistosomum haematobium* were randomly distributed in all stages throughout the testis. There were no zones and the cells were completely separated after divisions. He found no stage in which cells were interconnected, and he specifically stated that spermatids were never adherent to one another. It appears that in most schistosomes cytophores break apart easily if they are formed at all (Severinghaus, 1927), but in *Schistosomatium douthitti* the spermatids are connected to cytophores in groups of 32 (Nez, 1954). In general, trematodes not only show cytophores but all species which have been adequately studied appear to have three successive spermatogonial divisions resulting in cytophores with 32 spermatids with the exception of *Bucephalus* (Woodhead, 1931) in which only two spermatogonial divisions and spermatid groups of 16 were found (Table 8, p. 136). The degree of order within the testis varies widely even in related species. For instance, Chen (1937) found definite zones containing certain stages in *Paragonimus hellicotti*, but Willmott (1950) found random distribution of all stages in *Gigantocotyle bathycotyle*. In this species there is, however, a regular formation of cytophores which have the shape of plates because all cell divisions, three spermatogonial and two meiotic, occur in the same plane. In the turtle lung fluke, *Heronimus chelydrae*, Guilford (1955), who listed the pertinent literature quite completely, observed that primordial germ cells, and primary and secondary spermatogonia were irregularly distributed in clumps throughout the length of the testis. He described the formation of typical cytophores (Fig. 20) and the occurrence of masses of cells in degeneration at the spermatogonial, spermatocytic and spermatid stages. Irregular distribution of stages without zonation was also seen by John (1953) in *Fasciola hepatica*.

A great many detailed observations are to be found in two papers on the frog lung fluke, *Haematoloeachus medioplexus* (Burton, 1960, 1972), the first a light microscopic, the second an ultrastructural study. In this animal as in almost all

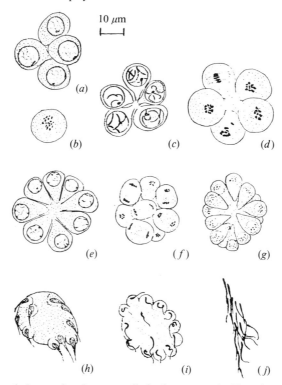

Fig. 20. Tracings of clones of male germ cells in the trematode *Heronimus chelydrae*. The development of a cytophore begins in spermatogonial stages (*a*) and continues through spermatocytes I (*c, d*) and II (*e, f*) to spermatids (*g, h, i*). (*j*) Free spermatozoa; (*b*) single spermatogonium in metaphase. (After H. G. Guilford (1955). *Trans. Am. microsc. Soc.* **74**, 182–90.)

flatworms the germ cells are small and do not stain easily. A thin layer of 'germinal cells' (primordials?) is in contact with the fibrous sheath of the paired testes. In the next layer are primary, secondary and tertiary spermatogonia which are not easily distinguished from the 'germinal cells'. All have about the same size but the later spermatogonia tend to be somewhat larger (germinal cells 4×6 μm, spermatogonia I and II 5.5–6 μm, spermatogonia III 7 μm). The most meaningful characteristic is the number of cells in a cluster. Spermatocytes grow only to a diameter of 8–9 μm. The rosette pattern of cytophores and deviations from the pattern in other trematodes are discussed. The spermatozoa vary remarkably in length from 200–483 μm with an average of 400 μm. The electron microscope shows much evidence of incomplete cell division resulting in the cytophoric rosettes (Burton, 1972). Spermatids remain connected at their basal ends by cytoplasmic bridges in clones of 32 cells. As spermatocytes and spermatids with long cytoplasmic tails are mingling in the viscous fluid of the testicular cavity it is difficult to identify stages accurately and to follow the complex process of spermiogenesis through all its steps.

In the Cestoda also cytophores have been found, for instance in *Moniezia* (Child, 1907), *Taenia* and *Rhyncobothrium* (Young, 1923), and *Hymenolepsis nana* and *H. diminuta* (Rosario, 1964). With the electron microscope Rosario identified the main cellular stages but was unable to find typical mitotic and meiotic morphology. He has made it plausible that the difficulty lies with the peculiar phenomena of cytophore formation in Cestodes, but the ideas of Child who believed that spermatid nuclei in this group originate through fragmentation, i.e. amitosis, and of Young who suggested that in meiosis the skeins of chromatin are scattered through the cytophore and reconstituted as nuclei of secondary spermatocytes, were not clarified or dispelled.

On superficial examination it appears that two issues of spermatogenesis in Platyhelminthes are quite clear. There are no nurse cells in the testes and cytophores arise through incomplete cell division, but neither one of these points is entirely beyond doubt. There are Dingler's observations on large cells, possibly nurse cells, among the spermatogonia of a trematode, and Sato, Oh & Sakoda (1967) in an electron microscopic study of the lung fluke, *Paragonimus miyazakii*, showed a picture (their fig. 1) of 'spermatogonia enclosed by a supporting cell' without further comment. Many questions remain with regard to origin and variations of cytophores. Some of these will be discussed in Chapter 11.

Nemertina

Very little is known about spermatogenesis in Nemertina. Some indications are given by Riepen (1933) who depicted a cross-section through a testicular follicle of *Malocobdella grossa*. A few large spermatogonia are situated, usually in pairs, in a simple lining epithelium and some spermatozoa are seen in a wide lumen. The mature testis according to Riepen is usually filled with a profusion of developing gametes in many different stages. There is no indication of cytophores or nurse cells. Humes (1941) observed that in the genus *Carcinomertes* the male possesses numerous oval follicles, only 27–37 μm in diameter, with a common efferent canal. In a single worm many different stages of spermatogenesis are seen, but in each 'testis' the gametes are synchronous. Up to several hundred spermatozoa were counted per 'testis', which may therefore represent only one or at best very few clones. Humes stated that all other known nemerteans have simple sac-like 'testes' with individual efferent ducts. Riser (1974) has reviewed reproduction in this phylum and has pointed out that classical histological methods are very defective in preserving the details of soft tissues in these very muscular creatures. The use of such methods has resulted in a confused literature and discouraged investigators.

Aschelminthes

The phylum Aschelminthes combines very diverse classes of which only the Nematoda have been extensively investigated with regard to spermatogenesis. The most complete review of the anatomy of nematode testes is by Chitwood & Chitwood (1940). With regard to spermatogenesis, *Ascaris megalocephala*, the intestinal roundworm of the horse, one of the most famous subjects of cytology, has been very extensively investigated ever since Van Beneden & Julin (1884). Hertwig (1890) set forth the advantages and disadvantages of the animal for research: each testis contains all stages from a primordial cell to a mature spermatozoon and does so continually throughout the year; the organ is a single tube several meters in length and on cross-section is usually cut 10–20 times, but each section represents only a single stage of spermatogenesis which may make the investigation somewhat laborious; also the spermatozoa are quite atypical and the earlier stages somewhat opaque because of their content of 'yolk granules' ('Dotterkörner'). Hertwig distinguished three testicular zones: zone I beginning at the blind distal end, a region of vigorous cell proliferation; zone II a region of growth in which the circumference of cells grows tenfold and refringent fat granules increase; and zone III in which spermatids arise through the spermatocytic divisions. In zone I the gametes do not lie free in the testis but are attached by fine threads to a central protoplasmic mass, the so-called "rachis". As germ cells multiply the rachis branches dichotomously and is seen to be connected with more and more cells. Through eight lamellae (branches of the third order) the rachis divides the testis contents into eight cords or strands of cells. At this stage of development the great majority of the germ cells (tertiary spermatogonia?) are attached to the tertiary ramification of the rachis. At the attenuated blind end of the testis five to eight cells are found which measure 25 μm in diameter; in the distal part of zone I there are more than ten times as many cells of diameters of 12–15 μm per cross-section. In between the cells 'Hodenzwischenkörperchen' are found (6 μm or smaller) which have a cytoplasmic and a chromatic part and which Hertwig considered degenerative cells. In zone II the spermatocytes grow two- to fourfold and become laden with 'yolk granules'.

Hertwig's paper, in which he discussed the main cytological arguments of the day and some which are still pertinent in our time, was excelled by a remarkable account, 323 pages long, and illustrated by photographs and frames from moving pictures, by Fauré-Fremiet (1913). His description is even richer in detail than Hertwig's, and his investigation included histochemical methods. In general, he confirmed Hertwig's findings. He concluded that the rachis constitutes 'the cytoplasmic remains of incomplete cell divisions'. The structure disappears in the zone of the spermatocytes. The granules found in the rachis and in the spermatogonia connected to it are neutral fat. In zone III four spermatids (diameter 15 μm) are associated with a cytophoric mass (4–5 μm) the origin of which has been described minutely by Van Beneden & Julin (1884). The cytophore of *Ascaris* is thus not the product of incomplete spermatogonial divisions as in platyhelminthes and other animals, but arises somehow in late spermatocytic or early spermatid stages. Prestage (1960) in an ultrastructural study of *A. lumbricoides* considered the rachis a syncytium formed by parts of spermatogonial cytoplasm, but he was unable to clarify its origin in detail, nor did he elucidate the development of the cytophore. Favard (1951) in another electron microscopic study concluded that spermatogonia, in contrast to oogonia, detach from the rachis after their last divisions. He then described the formation of the cytophore beginning in the spermatid before the second meiotic division is completed, through a stage where the cytophore appears as a cytoplasmic lobe containing almost all endoplasmic reticulum and some mitochondria, to its final separation and phagocytosis by the partial cells of the testis. None of Favard's pictures show unequivocally that more than one spermatid is attached to a cytophoric mass, nor do they demonstrate that this 'cytophore' is not something like a large residual body cast off from the spermatid as it occurs in many animals. Although almost all previous research has suggested that the spermatids in nematodes are connected, usually in quadruplets, the nature and origin of the connection is still not clear.

The granules which characterize all stages of male gametes in *Ascaris* have led to repeated histo- and cytochemical investigations. Fauré-Fremiet in the first of these, found that the brilliantly refringent granules which appear during the growth period of the spermatocytes consist of a proteinaceous substance he called 'ascaridine', which is free of phosphates and sulfates. In the spermatid these granules fuse to form the 'refringent cone' which later constitutes the tip of the ameboid, non-flagellar spermatozoon. Pasteels (1951) showed that in the spermatocytes RNA (pyronophilic substance) increases in proportion to the formation of ascaridine granules. Most of the RNA becomes concentrated in the cytophore and is finally lost from the spermatid. Favard has clarified this story morphologically. He demonstrated that the granular endoplasmic reticulum undergoes a remarkable growth in the spermatocytes and is the site of initial secretion of the proteinaceous substance subsequently packed into granules within the cisternae of the Golgi region. Many authors have pointed to the great similarity between the formation of these unique granules in the male gametes of *Ascaris* and the formation of proteinaceous yolk in oocytes of *Ascaris* and other animals. In another nematode, *Rhabditis belari*, Nigon & Delavault (1952) demonstrated massive RNA emission from the nucleus at the time of the maturation divisions, thus confirming incidental findings of similar phenomena in other worms (e.g. Cobb, 1925).

Fragmentary information often confusing in detail, exists for a great number of free-living nematodes. Walton (1924, 1940) has thoroughly reviewed the older literature. A particularly interesting and controversial case is the marine worm *Spirina parasitifera* (Cobb, 1925). According to Cobb, 'primordial spermatogonia' in this animal give rise by mitosis to 'twin cells' which move down the testicular tube in tandem. Each small spermatogonium grows into a primary spermatocyte (diameter 40 μm) which divides first at right-angles to the body axis and then parallel to it, and thus gives rise to four spermatids. 'Moving along the testis with soldier-like precision, the two caudal members form a tandem, followed by the other two, also in tandem; i.e. the quartet falls into single file. These spermatids in file grow and one after another divide internally without evidence of mitosis into 64 uninucleate elements which proceed to surround themselves with walls and *form a tissue of 64 cells*.' A mitotic division follows which results in 128 cells. Cobb stated that similar processes occur in other nematodes but that the aggregation into a 'tissue' is unique for *Spirina*. He created for a plurality of cells derived from a spermatid the term 'spermatidium', and such a group in the mature condition in which it is capable of fertilization, he called a 'spermule'. Walton stated in a footnote that according to Chitwood (no reference given) some of Cobb's observations may be in part erroneous. It must be pointed out, however, that divisions of spermatids have been reported in other animals, for instance in spiders and mammals, and should not be considered 'impossible' *per se*.

Triantophyllou & Hirschmann (1964) have reviewed reproduction in plant and soil nematodes, but found mainly that 'gametogenesis follows the general pattern known in most animals'. Hope (1974) in a review of marine nematodes described spermatogenesis in *Deontostoma californicum*. The testis is surrounded by an extremely thin tunica propria which Hope calls 'glycocalyx'. Under this lies an epithelium which the electron microscope shows to be non-syncytial. This encloses the germ cells which as in other marine nematodes are not attached to a rachis. The primary spermatocytes have very little cytoplasm and are 'definitely smaller than most spermatogonia'. At the proximal end of the testis the non-flagellated spermatozoa were found in an amorphous substance, probably a secretion, but it could not be excluded that it might consist of the ameboid processes of epithelial cells. These processes phagocytize the degenerating spermatozoa which occur frequently.

Taylor (1960) in the first investigation of spermatogenesis in filarial nematodes, *Litomosoides carinii* and *Dirofilaria immitis*, found in the single testis a proliferative and a growth zone. In the first, cell types were seen which were interpreted as nutritive cells filled with lipid droplets and spermatogonia with prominent nucleoli and granular cytoplasm. (A large spermatogonial nucleolus was described also in an ultrastructural study of *Rhabditis* by Beams & Sekhon, 1972.) Taylor found no rachis, and cytophores are not evident in the photographs taken with phase contrast and ultraviolet microscopy. In the growth zone the nutritive cells are relatively less frequent. The fact that in this worm as in many other nematodes the chromosomes persist in their individual shape in the spermatozoa (Walton, 1940) was demonstrated also with ultraviolet microscopy.

In summary, spermatogenesis in nematodes has some unusual and variable features. In *Ascaris*, the only genus which has been very thoroughly and frequently examined, the outstanding phenomena are the formation of a rachis in the spermatogonial stage, the development of a cytophore, not yet entirely clarified, connecting a quartet of spermatids, and the unusual accumulation of proteinaceous 'yolk granules' in the spermatocytes. In many other nematodes these features have not been seen. A rachis is not a general feature of the class, but it appears that in many species in which a rachis does not develop, nurse cells of some kind play a prominent role. Cytophores are often absent and it seems uncertain whether the germ cells develop in clones. In all nematodes the unusual features of the spermatozoa are associated with a variety of unusual details in spermiogenesis.

The few investigations of the male gonads of the aschelminthic class of Rotifera have indicated some unique special features which deserve much finer analysis in the future. Tauson (1926) examined *Asplanchna intermedia* in which the early spermatogenic stages take place in the embryo. He stated that the first division of a spermatogonium is a reduction division which is followed by a series of seven or eight postreductional divisions. If these technically difficult observations are correct, then the resulting cells should be called secondary spermatocytes or, perhaps, spermatids. The finding is unique and does not fit well into the accepted picture of animal spermatogenesis but reminds one somewhat of the so called 'intermediate' or 'sporic' form of meiosis characteristic of higher plants and some thallophytes (Wilson, 1925), in which the immediate products of meiosis are not gametes but asexual haploid tissues, embryo sacs, pollen grains, etc.

Two kinds of sperm have been described in Rotifera (Whitney, 1917). These were studied with the electron microscope by Koehler (1965) and Koehler & Birky (1966). According to them, the atypical 'rod spermatozoa' are not cells but cellular products filled with microtubules. These are produced by atypical spermatids which later degenerate. The function of the rods is not known, but they are released together with the typical spermatozoa which are motile with a 'leading flagellum' and an undulating membrane. While nurse cells have not been observed in rotifer testes, Koehler has demonstrated that the early spermatids are intercellularly connected in groups.

Polyzoa

Some clear descriptive information exists for the phylum Polyzoa (Bryozoa or Ectoprocta), although it has been stated that spermatogenesis is difficult to investigate because the elements are usually small and the material 'little favorable and little spectacular' (Grellet, 1957), and only very few species have been examined. Marcus (1934) described the origin of cytophores through sloughing of the cytoplasm of spermatids and insisted that this structure in Polyzoa is neither comparable to Sertoli cells nor does it represent a nutritional device at all. We will find that later observations make this opinion seem at least arguable. As an incidental note of interest, Marcus mentioned that the microsporidium *Nosema bryozoides* lodges specifically in the spermatogonia of the bryozoan *Lophopus crystallinus*.

Grellet (1957) described the 'testicles' and spermatogenesis in *Alcyonidium gelatinosum*. The terms "testis" and "testicle" have been used traditionally for

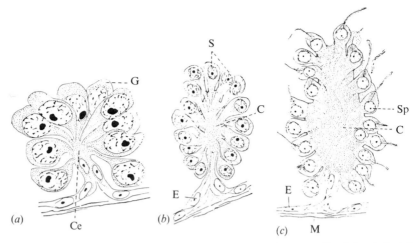

Fig. 21. Semi-diagrammatic representations of clones of male germ cells of the phylactolae-matous polyzoon *Plumatella fungosa*. (*a*) Spermatogonial; (*b*) spermatocytic; (*c*) spermatid stage. Ce, center of clone, early cytophore; G, spermatogonia; C, cytophore; S, spermato-cytes; Sp, spermatid; E, modified coelomic epithelium; M, myoid layer. (After P. Brien (1966). *Biologie de la Reproduction Animale.* Paris: Masson et Cie.)

the accumulations of germ cells in polyzoans (Hyman, 1959), although the so-called gonads are unencapsulated aggregations of germ cells derived from the peritoneum and in later developmental stages exposed to the coelomic fluid (Plates 11 and 12). In young zooids (all Polyzoa are colonial organisms) primordial germ cells, easily distinguishable from somatic cells, appear first in the coelomic lining. The early spermatogonia which somehow develop from the primordial cells are small (nucleus 2–4 μm, cell 3–5 μm) and normally divide mitotically, the cytoplasm cleav-ing completely or nearly completely, but Grellet occasionally observed nuclear division without cytoplasmic separation. In a number of successive divisions a mulberry-shaped, solid aggregate is produced in which 10–20 cells were counted in the spermatid stage. Grellet did not mention a stage between spermatogonia and spermatids. All cells of a 'morula' develop simultaneously. The spermatid morula forms a 'residual body' in which granules of RNA and lipid droplets were demonstrated. The spermatids remain adherent to this body for a considerable time. Brien (1960, 1966) has described and clearly depicted spermatogenic stages in both classes of Polyzoa, the Phylactolaemata and the Gymnolaemata. In a member of the first, *Plumatella fungosa* (Fig. 21 *a*, *b*, and *c*), spermatogenesis takes place on the dorsoventral septum of the common cavity of the colony. The spermatogonia appear among the epithelial cells lining the coelom and as they develop bulge out in bouquets on stems formed by the peritoneum. Each bouquet represents a 'follicle' and presumably a clone. The tips of the elongating sperma-togonia are fused into a cytophore and it is to this that the epithelial cells are attached. As spermatogenesis proceeds synchronously in all the cells of a follicle, the central cytophore grows in size and develops cytoplasmic processes which extend as narrow envelopes around the individual spermatids (Plate 12). In most Polyzoa the flagella show vigorous motility while the spermatids are still firmly attached to the cytophore (Silen, 1966, and personal observations on *Bugula bugula* and a species of *Membranopora*, Plates 11 and 12). The spermatozoa are finally released into the coelomic cavity. From Brien's account (1960) it appears that in some Gymnolaemata rosettes of spermatocytes and spermatids in different stages attached to a central cytophore are released into the cavity and presumably develop as free-floating cellular aggregates.

Phoronida

Almost all information available on spermatogenesis in Phoronida appears to be contained in a paper by Rattenburg (1953) on reproduction in *Phoronopsis viridis*. In general, she confirmed the earlier findings of Ikeda (1903). The gonads form around capillaries directly under the peritoneal lining. First a 'nutriment layer' develops, then the remaining, small and flat peritoneal cells begin to proliferate and form a 'crowded assemblage' of small cells which Ikeda called spermatogonia. Rattenburg did not feel that spermatogonia were ever identified satisfactorily. The nutriment layer is resorbed as spermatogenesis proceeds and connective tissue fibers penetrate the germ cell congregation. The gametes arrange themselves about the radial fiber bundles and form 'testicular strings'. The oldest stages are found at the ends of the strings. The heads of the spermatids point toward the central string but do not appear in bundles. Rattenburg described the spermatocytes as extremely small. Spermatids occurred in attached pairs. She was unable to identify any germ cells or reproductive tissue except in the annual breeding season.

In *Sipuncula* spermatocytes break off from the testis in loosely associated groups (cytophores?) which grow and develop in the coelomic cavity (Rice, 1974).

Echiura

The worm *Urechis caupo* is the only species of the small phylum Echiura which has been studied in some detail. The first extensive account of reproduction in this animal was given by Newby (1940) who was unable to find a gonad, but described the germ cells in all stages of development floating together with hemocytes and other ameboid cells in the coelomic cavity at all times of the year. The germ cells occur in primary and secondary aggregates. The primary ones contain 40–50 cells presumably of the less mature stages. In each cluster the cells are of similar size, in different clusters they range in diameter from 3.5–10 μm. The smallest of these, with relatively little cytoplasm and condensed chromatin, Newby considered to be spermatids. During spermiogenesis six to ten primary clumps gather into secondary ones which are surrounded by a delicately fibrous mass, probably formed, according to Newby, of sperm flagella and discarded cytoplasm. The secondary clumps break up before the spermatozoa are gathered in storage organs. On the average, a mature worm has nearly three billion sperm in these organs.

In labeling experiments with [³H]uridine and [³H]thymidine Das (1968) was able to clarify the story further. Through his observations it has become almost certain that spermatogenesis in *Urechis* begins in the coelomic cavity in early spermatogonial stages, perhaps with single spermatogonia. It is not certain how often the spermatogonia divide although the size of spermatocytic groups indicates four to six divisions. The primary spermatocytes do not show an appreciable growth. With the increase in chromatin condensation RNA synthesis decreases sharply. These results are in contrast to those found, for instance, in the spermatocytes of arthropods or other animals which show conspicuous growth of spermatocytes. Synthesis of proteins continues to occur, although at a decreasing rate, in stages in which RNA production is low or zero. Das & Alfert (1968) showed that histones are produced in spermatocytes of *Urechis* during the S-phase of the cell cycle. The [³H]thymidine label permitted an estimate of the duration of the period from the beginning of meiosis to the end of spermiogenesis of about 10 days at 12–14 °C. In general, spermatogenesis in *Urechis* is similar to that of many annelids.

Annelida

The development of the male gametes in the annelids has been quite thoroughly investigated in the oligochaete group of the so-called 'earthworms' which comprise possibly 1800 species (Stephenson, 1930) and of which 'any piece of lawn in New York City may carry as many as four or five species'. Oligochaete reproduction has been discussed in several reviews, e.g. by Beddard (1895) and Stephenson (1930). From a survey of nearly 100 years of investigation (beginning with the paper by Bloomfield, 1880) it appears that in annelids in general, spermatogenesis occurs to a greater or lesser extent in the open coelomic cavity or in more or less specialized coelomic recesses, and that these conditions may be classified from the simple to the complex regardless of any phylogenetic order within the phylum (Table 5). In the most primitive condition no testes exist and stem cells or early spermatogonia are released into the coelom. In slightly more complex cases the germ cells are retained in a germinal or gonadal tissue until spermatocytic stages. The greatest morphological complexity arises when special pockets of the coelom serve as sites for the later phases of spermatogenesis. The identification of these special organs has led to much confusion of terminology. The issue was first discussed clearly by Beddard. He correctly insisted that the coelomic outpouchings which form a secondary environment for the gametes released from germinal tissues should not be called 'testes', a name to be reserved for primary sites of gametic development. Beddard named the structures 'sperm-sacs' and rejected the term 'seminal vesicles' because it is pre-empted in vertebrates by structures

TABLE 5. *The location of stages of spermatogenic cells in various annelids*

Location	Class and genus
Stem cells or primary spermatogonia released into open coelom. No testes present.	Oligochaeta: *Aelosoma* (Bunke, 1967) *Chaetogaster* (Stephenson, 1922) *Enchytraeus* (Beddard, 1895)
Late spermatogonia or primary spermatocytes released into open coelom. Gonad more or less well defined.	Polychaeta: *Harmothoe* (Daly, 1974) *Arenicola* (Olive, 1972a, b) Archiannelida: *Saccocirrus* (Buchner, 1914)
Spermatogonia released from testis into testis-sacs.	Oligochaeta: *Tubifex* (Dixon, 1915)
Spermatocytes I released from testis into sperm-sacs.	Oligochaeta: *Lumbricus* (Anderson *et al.*, 1967)

of the same name whose derivation and function is completely different. Stephenson preferred 'testis sacs' which he defined as 'compartments of the coelom in which the testes lie'. He applied the term 'seminal vesicles' to a third type of receptacle which in some species, e.g. in *Lumbricus*, receives the germ cells in the last stages of their development. In the annelids, in which all stages of testis formation appear, from the complete absence of definable gonadal tissue to a pair of well defined testes as in the majority of oligochaetes, the definition of a testis must be carefully considered. The best definition is possibly "that circumscribed or demarcated tissue in which the earliest phases of spermatogenesis take place".

In the oligochaete *Chaetogaster orientalis* (Stephenson, 1922) germ cells proliferate diffusely in many body segments directly under the coelomic lining on strands between alimentary canal and body wall, or on the inner surface of the body wall and form a cellular tissue, the 'budding zone' which is also the source of new segments. Similarly, in many polychaetes the germ cells originate in peritoneal proliferations over large areas on the body wall, on blood vessels or on the alimentary canal (Daly, 1974). By far the most revealing description of the early development of germ cells was given by Salensky (1907) for an archiannelid species, *Protodrilus flavocapitatus*. Salensky stained the young animals supravitally with ammoniacal carmine, then cleared the specimens and stained with hemalum. The carmine was found in the germ cells located on thickenings of the somatopleure, the genital ridges, ventral of the gut. Here they are directly supplied by the ventral blood sinus and are also more or less exposed to the coelomic fluid into which they penetrate through the coelomic lining to about half of their circumference. Carmine granules were observed within the genital ridges and in the germ cells as long as they were attached. The granules were seen to extend

Fig. 22. Bipartite spermatogonial clone (32 cells) of *Lumbricus herculeus*. The two cytophoric centers are connected by a thin cytoplasmic band stretched in the preparation. (After E. Chatton & O. Tuzet (1941). *C. R. Acad. Sci., Paris* **213**, 373–6.)

in strands toward and into fine processes of the germ cells. Detached cells did not take up the dye. According to Salensky, as soon as the first free germ cells are being produced the peritoneal epithelium begins to proliferate into a kind of parenchyma ('coelenchyma') which surrounds the germ cells from all sides. Salensky's definition of the gonads is not very clear. He described their origin as follows: 'After the primordial germ cells in the female and in the male have reached a considerable size, they begin to divide and form groups on both sides of the ventral sinus which represent ovaries or testes. These gradually fill the whole cavity of the segment and display as is usually the case with gonads, a whole series of different stages of development of the germ cells'. Salensky suggested that each newly forming segment of the worm becomes populated with germ cells from the preceding segment through the genital ridge and the hemolymph.

In the Annelida the developing sperm cells whether they are found in the coelom, in the testes, or in the sperm-sacs are always seen in groups, often called 'rosettes', around a central anuclear mass of cytoplasm, the so-called 'cytophore'. This structure has been described above for the Platyhelminthes and Polyzoa. Its origin in annelids has been often discussed (Bloomfield, 1880; Voigt, 1885; Bugnion & Popoff, 1905; Depdolla, 1906; Chatton & Tuzet, 1943; Walsh, 1954; and many others) and there is general consensus that it arises in early spermatogonial stages. It is presumably a modification of incomplete cell division.

Chatton & Tuzet (1943) found it impossible to determine in *Lumbricus herculeus* whether the primary spermatogonia separated completely after their division, but it was evident in their preparations that the cells in subsequent divisions remained connected by strands of cytoplasm. By careful study of squashes and of preparations briefly fixed in osmium vapors and stained with hematoxylin, the authors found that a 32-cell morula of spermatogonia consists of two parts with 16 cells each. In each half the cells have a common cytoplasmic center to which they adhere by thin strands. The two centers are connected to each other by a similar cytoplasmic strand (Fig. 22). The cytophore, therefore, has a complex configuration. In addition, it presents many pseudopodia which may be long and ramified, and which were observed to be highly motile in life. The composition of the cytophore as two centers connected by a thin interconnection appears to lead often, but by no means always, to a fission of the 64-spermatocyte groups into two of 32 cells each. The result is that aggregates of 256 spermatids are considerably more rare than aggregates of 64 primary spermatocytes, although they would be expected to be more frequent.

In *Lumbricus terrestris* the zone of attachment between spermatids and cytophore appears as 'a dense cytoplasmic band from which 20–30 Å filaments arise' and mitochondria and an elaborate endoplasmic reticulum occur in the cytophore (Anderson *et al.*, 1967). Electron microscopic investigations of another earthworm, *Eisenia foetidus*, have revealed that the connecting bridges between cells and cytophore are covered with a 'fuzzy coat'. In the spermatid stage the cytophore contains annulate lamellae, degenerate dictyosomes and cisternae of endoplasmic reticulum, presumably received from the perinuclear cytoplasm of the germ cells (Stang-Voss, 1970). This is of interest, because it seems to contradict the 'simple' role of the cytophore as a center of nourishment, typically represented by Depdolla (1906). In an electron microscopic study of *Branchiobdella pentadonta* Bondi & Farnesi (1976) found that meiosis begins in 'follicles' containing 16 spermatogonia connected to a cytophore (four spermatogonial divisions). The spermatids separate from the cytophore only when their differentiation is complete. Bondi & Farnesi saw synaptonemal complexes in primary spermatocytes. They also gave a detailed description of the spermatozoon of *Branchiobdella*.

There is one notable exception to the fairly consistent pattern of spermatogenesis in annelids in the species *Myzostomum cirriferum* (an animal which is considered an aberrant polychaete). Jägersten (1934) has given a detailed account of the male reproductive organs of this animal mainly from observations of living or fresh material. If this animal can indeed be considered an annelid, it is the only annelid reported to have a testis filled with spermatocysts very similar to those of arthropods. The cysts contain usually 8 or 16, but occasionally about 30 cells. When a fresh cyst is squashed the cells easily separate but in the spermatid stage the cells adhere to each other by a central anuclear cytoplasmic mass, a cytophore. Jägersten suggested that this might arise through the adhesion of cytoplasmic lobes pinched off from the individual spermatids.

Chaetognatha

Knowledge of the reproductive processes in Chaetognatha goes back to Hertwig (1880) and Lee (1887), and what little is known about spermatogenesis was summed up by Hyman (1959). A pair of ribbon-like testes lie closely applied to the lateral walls of the anterior tail coelom. According to Hyman spermatogonia are budded off in masses into the coelom where spermatogenesis takes place while the cells drift in a circulating coelomic current. The spermatozoa are funneled into so-called 'seminal vesicles' where they are combined into spermatophores.

Pogonophora

The Pogonophora (Ivanov, 1963) have a pair of seminiferous sacs or testes lined with an epithelium which appears to be a continuation of the splanchnopleure. Anteriorly this is well developed and ciliated, posteriorly it is represented by a simple layer of large transparent cells without cilia. In the lumen, which presumably is an extension of the coelom, all stages of spermatogenesis are represented, most of them in the form of characteristic morulae. There are large single cells (possibly stem cells), quadruplets of spermatogonia, morulae of spermatocytes and masses of filiform spermatozoa. As in platyhelminthes and annelids the morulae represent anuclear central cytoplasmic masses (cytophores) to which the germ cells are connected. Dimorphism of spermatozoa occurs (Franzén, 1973).

Echinodermata

Echinoderms in the state of reproductive activity have voluminous, arborescent, lobulated or tubular testes surrounded by a muscle layer which is directly involved in the process of spermiation. According to Hyman (1955) in Holothuroidea, which are usually dioecious but occasionally hermaphroditic, a single gonad is located in the anterior part of the coelom. The testis usually consists of numerous tubules often of considerable length (in *Stichopus japonicus* 25–35 cm during the reproductive season). Asteroidea typically possess 10 gonads, two in each arm, but often the testes occur in great numbers serially aligned along the side of each arm. In Echinoidea, which are strictly dioecious, the number of testes is typically five; in Ophiuroidea they vary from one large one to thousands of small ones.

It is astonishing that spermatogenesis in echinoderms has not been more thoroughly explored in spite of the great attention which cytologists and embryologists have devoted to the eggs of sea-urchins. Only scattered information is available. The gross features of the reproductive cycle and details of the testicular envelopes (Wilson, 1940; Tangapregassom & Delavault, 1967; Walker, 1974) have been studied repeatedly, but the spermatogenic process in its continuity has not been thoroughly investigated in even a single species. Whatever evidence is available comes from the class Echinoidea and in particular from the order

Echinoida. Caullery (1911) described for *Echinocardium cordatum* an outstanding feature, now established as common in echinoderms (Cognetti & Delavault, 1958). In *Echinocardium* during the period of sexual inactivity following a reproductive season, the testes are almost entirely composed of 'large cells, each of which contains a vacuole and numerous spherules of reserve substance'. 'Agglutinated', i.e. degenerating, spermatozoa stick in pockets apparently ingested by the large cells. Small groups of spermatogonia are found around the periphery of the sac-like organ. This is the state of the gonads for at least half of the year (July to December in Wimereux, France) but even at the height of maturity in April and May the large cells never disappear completely. According to Holland & Giese (1965) who have investigated the cells thoroughly, the most appropriate name for them is 'nutritive phagocytes'. They occur in the ovary as well as in the testis, and alternate between two main phases, a globulated and a deglobulated one. In *Strongylocentrotus purpuratus* a nutritive phagocyte in the globulated phase is spherical or oval with an average diameter of 35 μm and contains large spherical inclusions. The nucleus is small (3×4 μm). It is obvious in light microscopic preparations that the globules contain ingested cells and cellular debris and are, therefore, phagosomes. At the climax of the deglobulated state, which coincides with spawning and the immediately following period, the phagocytes measure approximately 15 μm in diameter and are devoid of globules or vacuoles. The cells become difficult to detect and reports of their total disappearance may be due to this feature. Phagocytes may be labeled with [3H]thymidine during the deglobulated phase (Holland & Giese) and apparently give rise to phagocytes which may be found labeled in the next annual cycle. The process of spermatogenesis in *Strongylocentrotus* is best summarized in a brief synopsis of the stages found by Holland & Giese:

(1) Spring (April, May). After spawning, phagocytes are deglobulated and vacuolated. Leftover spermatozoa are massed in the acinar lumen and also scattered throughout all levels of the germinal layer often within phagocytes. Spermatogonia ('primitive') are widely scattered in the most basal layer. Some are labeled after 1 hour's exposure to [3H]thymidine.

(2) Summer (July, August). Phagocytes contain many globules and relict spermatozoa. Some spermatozoa are still in the acinar lumen. Germ cell nests at the base of the germinal tissue contain 'primitive' and secondary (smaller) spermatogonia which may be labeled. In August the nests fuse and form an ever-thickening continuous band of spermatogonia (peripheral) and spermatocytes (central). Spermatocytes do not gain appreciably in volume. Spermatids appear in radial groups, 'trails', between adjacent nutritive phagocytes. Spermatids appear labeled 6 days after exposure to [3H]thymidine. In September, the picture changes only quantitatively. Small masses of spermatozoa appear in the lumen. Forty days after administration of label spermatozoa are labeled in the lumen and in clumps ingested by phagocytes.

(3) (October, November). Labeled spermatozoa are found only in the lumen and not in the phagocytes.

(4) (December, January). The lumen is distended by huge masses of spermatozoa. The phagocytes are nearly completely deglobulated and have ceased to ingest. The germinal layer is thinner but 1 hour's exposure to [³H]thymidine labels many 'primitive' and secondary spermatogonia and spermatocytes. After spawning (January) only few primitive spermatogonia show label.

(5) (February, March). Phagocytes remain deglobulated. Spermatogonia and spermatocytes still label but become detached toward the lumen in large clumps. Labeled spermatozoa produced probably *after* spawning, occur in the lumen.

This annual cycle typical of many echinoderms indicates several characteristic points which need further exploration. Spermatozoa are generated and differentiating at an even pace (the duration of spermiogenesis remains constant) for many months and a large proportion of them are stored in the voluminous testicular cavity until spermiation. Others degenerate and are at least in part ingested by the nutritive phagocytes which after spermiation remove the residual spermatozoa. Spermatogonia and spermatocytes divide throughout most of the year, even after spawning when most of them abort except for primary spermatogonia. Although the configuration of strands of spermatids suggests clonal development and ultrastructural evidence shows that spermatids are connected by intercellular bridges (Longo & Anderson, 1969), there is little information on the number of germ cells in a clone and the nature of cellular interconnections within it. Order in the germinal tissue is restricted to the peripheral position of young and the central position of older stages. The population of phagocytes has an annual cycle of behavior correlated with the cycle of germ cells. At spawning time when phagocytic activity is at its lowest, many of the phagocytes are apparently extruded with the sperm, but new ones are formed through division of the ones which remain.

A somewhat different although similar story has been presented for *Stylocidaris affinis*, another sea-urchin (Holland, 1967). In this animal the spermatocytes accumulate in the germinal tissue throughout winter, spring and summer in a continuously thickening layer. During all this time meiosis is not completed. In the months remaining before spawning, meiosis occurs and spermatids and spermatozoa differentiate. The behavior of the nutritive phagocytes is very similar to that in *Strongylocentrotus*. The seasonal cycles of other echinoderms such as holothurians (Tanaka, 1958) and ophiuroids (Patent, 1969) indicate that the main features of spermatogenesis are similar.

An ultrastructural analysis of the nutritive phagocytes throughout their life cycle is most desirable. In a study on spermiogenesis in *Arbacia punctulata* and *Strongylocentrotus purpuratus*, Longo & Anderson (1969) described the fine structure of what they called 'interstitial cells' which they found 'mixed with the developing sperm cells' and which are almost certainly the phagocytes described by Holland and others. The interstitial cells contained PAS-positive droplets, lipid bodies and cytoplasmic masses similar to the discarded cytoplasm of late spermatids. They possessed a flagellum with a large kinetosome. It is obvious from the context that these findings are restricted to a very small part of the behavioral cycle of the phagocytes.

The spermatogonia, spermatocytes and spermatids of sea cucumbers (Holothuroidea) have been studied recently with the electron microscope (Atwood, 1974). Unexpectedly, the spermatogonia which usually are applied to a basement membrane at the testicular wall are not connected by cytoplasmic bridges but by 'desmosome-like' structures. They are not organized into a definite layer or in groups, but lie scattered among the spermatocytes. The spermatogonia in *Cucumaria lubrica* occasionally contain a crystalloid tubular body which is similar to the crystalloids of Lubarsch in the human (Nagano, 1969).

According to Atwood spermatocytes in Holothuroidea show various degrees of flagellar formation. Spermatids are joined by intercellular bridges. The maximum number of cells seen interconnected was three. Atwood came to the tentative conclusion that only one mitochondrion is present throughout spermatogenesis which transforms from a tubular, branched structure in the early spermatid into a compact rounded body in the spermatozoon.

8

Vertebrata

The vertebrates are a relatively closely related, homogeneous group of animals, and the basic structure of their gonads and their seminiferous tissue are more readily comparable with each other than with the invertebrates. Homologies are usually obvious and easily documented. Many of the facts about vertebrate reproduction are well known to zoologists and anatomists and accessible in standard works of morphology and physiology. Mammalian spermatogenesis has received much attention in the past 20 years and has been repeatedly and quite thoroughly reviewed. There is, therefore, less need for a searching review in the vertebrates than in the invertebrates. However, it is necessary to establish an outline of the main features of spermatogenesis in this subphylum in the interest of a systematic comparison of all animals. The lower vertebrates will receive more detailed treatment than the mammals which will be presented essentially as a review of reviews.

It is difficult to characterize a single type of vertebrate spermatogenesis. Rather there are two types, one in fishes and amphibians, and one in reptiles, birds and mammals, i.e. amniotes. In fishes and amphibians the development of spermatogenic clones proceeds within cysts which are located within tubular compartments called lobules or tubules. In amniotes cysts are absent, and seminiferous tubules contain the germinal tissue without subcompartments. In both types the supporting or Sertoli cells have characteristically intimate structural association with the germ cells. It appears also that the nervous and endocrine controls superimposed on spermatogenesis in vertebrates are typically more complex than in other animals.

The gonads of vertebrates are relatively compact in contrast to the often multiple or arborescent sex organs of many invertebrates. In large species the testes are of considerable size and have a complex tissue framework which separates and subdivides lobes, lobules and primary seminiferous compartments. Endocrine secretory cells are situated characteristically, although probably not in every case, in the interstitial tissues. Typically the testes are enveloped by a strong tunic of connective tissue continuous with septa coursing toward the mesenteric connection of the gonad with the body wall. The tunic frequently contains myoid and occasionally endocrine cells. The germinal tissue is contained in primary compartments in the shape of sacs or tubules often connected to the efferent ducts by a short tubular link. In mammals the seminiferous tubules are essentially loops which empty their contents through both limbs. In lower vertebrates the compartments are more or less spherical "follicles" or "ampullae" and may be closed except at the time of spermiation.

Fish

Marsipobranchii

There are no detailed accounts of the process of spermatogenesis in lampreys and hagfishes but the origin of their gonads and the early development of their germinal tissues have been unusually well investigated (Cunningham, 1886; Okkelberg, 1921; Butcher, 1929; Hardisty, 1965*a*, *b*). Although embryology is not a central theme of this book, Hardisty's findings are a most suitable introduction for a discussion of gonadal structures in vertebrates, and will, therefore, be reviewed here briefly without explicit reference to the older literature which is well reviewed in Hardisty's papers.

The gonads in the ammocoete larvae of lampreys go through an early female phase of development regardless of their definitive sex determination. According to Hardisty, the 'primordial germ cells' in the young larva migrate out of the caudal entoderm to positions under the peritoneal epithelium covering the gonadal ridges on the dorsal body wall (Hardisty did not use the term "germinal epithelium"). They may lie among the epithelial cells and even appear to project into the coelom, but 'in especially favorable cases it may be seen that their outer surface is covered by a thin membrane representing the cytoplasmic extension of the peritoneal surface cells'. The products of primordial cell divisions are 'protogonia', similar to their mother cells but considerably smaller. Each protogonium acquires its own 'follicular' envelope. When two or more cells are seen enclosed in a follicle they are defined as 'deuterogonia' or secondary spermatogonia. In deuterogonial cysts germ cells develop synchronously. Meiosis and oocyte growth in the larva occur in large cysts containing several hundred cells which may advance as far as the diplotene stage but degenerate eventually. Regardless of the definitive sex of the animal the female phase may persist for several years. Hardisty suggested that there is 'a continuing feminizing influence on the undifferentiated germ cells, the intensity of which decreases as the morphological differentiation of a testis progresses'. In the development of a male the somatic tissues in the gonad indicate the sex earlier and more reliably than the germ cells. A dense mass of fibrous connective tissue fans out from the gonadal mesentery and blends with a rudimentary tunica albuginea under the peritoneal epithelium. There is no evidence of a migration of germ cells from the cortical parts of the testis to the deeper regions. The male deuterogonia in their follicles apparently are maintained for periods of years before spermatogenesis begins. Unfortunately, Hardisty has given no description of later stages of development or of the adult testis, but his discussion of mechanisms of sex determination in vertebrates and invertebrates is exemplary.

Selachii

The histology of the selachian testis is an attractive subject (Plate 13). The gametes are comparatively large and form conspicuous developmental patterns. Valette St George (1878) began with a thorough exploration of selachians. Jensen (1883) investigated spermatogenesis in a ray and compared it with invertebrates, and Swaen & Masquelin (1883) compared selachians with other vertebrates. However, the results of the older investigators are difficult to interpret because the terminology of developmental stages had not been established at the time. A much later account (Matthews, 1950) is particularly illuminating and excellently illustrated (Plate 13), although it contains some significant errors. According to Matthews, the large testis (18×37 cm) of the basking shark *Cetorhinus maximus*, consists of many roughly spherical lobes made up of wedge-shaped lobules whose apices meet at the center where numerous nests of cells lie in a strand of connective tissue. Each nest develops from a primary spermatogonium enveloped by flattened auxiliary cells. As the spermatogonia multiply a lumen develops and the cell nests become 'ampullae' which move outward into the lobules as spermatogenesis proceeds. According to Matthews the multiplication of germ cells at first results in the formation of a dense layer of spermatogonia around the central lumen and several layers of spermatocytes constituting the bulk of the ampullar wall. The spermatogonia then migrate to the periphery where they 'become Sertoli cells' (Plate 13, *a*, *b* and *c*). This is without doubt an error of observation and interpretation. Stanley (1966) has pointed out that the cells which Matthews called spermatogonia probably are Sertoli cells which have been shown to undergo just this shift in position in other selachians (Fig. 23). Subsequent stages of spermatogenesis in the basking shark, particularly the formation of spermatid bundles and their close association with Sertoli cells are seen in Plate 13(*e*) and (*f*). Stanley's investigation permits a comparison between *Cetorhinus* and *Scyliorhinus caniculus*, both members of the order Pleurotremata, and *Torpedo marmorata* of the Hypotremata. In general, the patterns are very similar, but while in *Cetorhinus* the germinal sites are located centrally in the testis and move outwards as they mature, in *Scyliorhinus* and *Torpedo* they originate on the lateral or dorsolateral aspect of the rostral end of the testis and move along the gonadal axis medially and caudally. According to Stanley, an early follicle in shark or ray contains one or two spermatogonia and is surrounded by several 'epithelial' supporting cells considered homologues of 'Sertoli cells' in mammals (see Chapter 11). These cells divide until there are 480–500 per follicle in *Scyliorhinus* and 230–260 in *Torpedo*. At this time follicle size is approximately 100 μm in both species, but in *Torpedo* germ and supporting cells lie intermingled, while in *Scyliorhinus* they are in separate layers. Spermatogonia in the course of their divisions undergo nuclear changes very similar to those seen in the rat when the cells differentiate from 'dusty' to 'crusty' nuclei (Regaud, 1901) or from 'Type A' to 'Type B' (Allen, 1918). Four successive spermatogonial divisions probably take place within the follicle as indicated by the number of

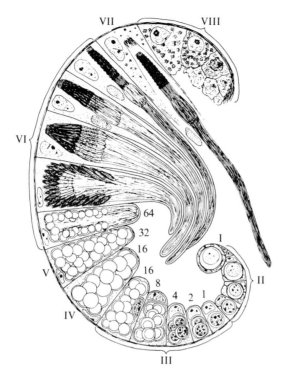

Fig. 23. Diagrammatic representation of development of spermatocyst in the shark *Scylio-rhinus caniculus*. I, newly formed cyst; II, spermatogonial proliferation; III, continuation of spermatogonial proliferation. Now each clone is enveloped by a cyst cell. Arabic numerals indicate number of germ cells in clone. IV, spermatocytes I; V, spermatocytes II; VI, spermiogenesis; VII, spermiation; VIII, degeneration of cyst. Each lobule or "follicle" contains many spermatocysts in the same stage of development. (After H. P. Stanley (1966). *Z. Zellforsch. mikrosk. Anat.* **75**, 453–68.)

spermatids (64) attached to a single supporting cell, which was found by Stanley in six different selachian species. In each follicle germ cells develop synchronously. In the spermatocytic stage the germ cells grow so strongly that the size of the follicle doubles. At the time of the meiotic divisions spaces begin to develop between the germ cells. Each supporting cell forms a separate pouch containing a clone of spermatids. This unit is the "spermatocyst" (Fig. 23). Cytoplasmic bridges are seen between many of the spermatocytes, a fact which has been confirmed in electron micrographs.

The key to an understanding of the process of spermatogenesis in elasmo-branchs lies in the behavior of the supporting cells which Holstein (1969) has greatly clarified in a study on *Squalus acanthias*. In this animal the developing ampullae migrate, through the length of the testis as they probably do in most selachians with relatively small testes, so that successive zones of ampullae are formed. The rostral zones contain the spermatogonial stages, the caudal zones the

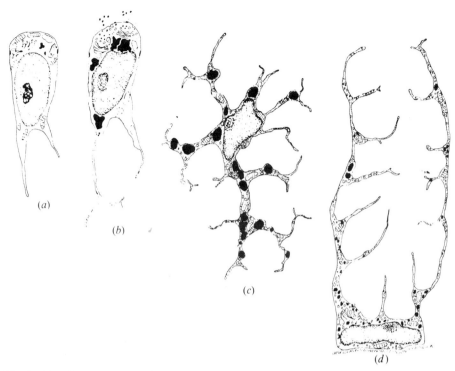

Fig. 24. Development of the cyst cells of *Squalus acanthias*. 'Prismatic' cell (*a*) develops into 'secreting cell' (*b*) during spermatogonial stage. During meiosis the nucleus migrates to the periphery of the ampulla (*c*) and the cell becomes a 'nurse cell' surrounding the germ cells with its processes (*d*). (After A. F. Holstein (1969). *Z. Zellforsch. mikrosk. Anat.* **93**, 265–81.)

spermatids. According to Holstein, the history of the supporting cells is as follows. When ampullae form from primitive cell nests the spermatogonia are of a secondary type (B) and the early lumen is lined with 'prismatic' epithelial cells (not with spermatogonia as described by Matthews). At this stage the epithelial cells show in electron micrographs rows of mitochondria under a smooth luminal surface and somewhat later there are signs of secretion of osmiophilic droplets by exocytosis into the lumen. Thin cytoplasmic processes are seen to penetrate between the multiplying spermatogonia (Fig. 24). When an ampulla has acquired eight layers of germ cells the epithelial cells leave the lumen and move between the spermatogonia to the periphery, their processes enveloping a certain number of germ cells. The cytoplasm of the migrating cells becomes more electron-dense and contains numerous lipid droplets. The germ cells begin to demonstrate signs of early meiosis. When the former 'prismatic cell' has reached the periphery it is transformed into a 'nurse cell' which Holstein regarded as analogous (not necessarily homologous) to the mammalian Sertoli cell. Its nucleus is basally situated and the cytoplasmic processes contain as in a cup a group of 64 spermatids which transforms into a fascicle of spermatozoa. The number of nurse cells is the same as the number of spermatocysts or, when mature, bundles of spermatozoa, on the average 424 per ampulla. The number of spermatozoa per cyst indicates that four spermatogonial cell divisions occur after a cyst is formed. From the fact that there are 424 fascicles per ampulla, Holstein concluded that approximately 10 spermatogonial divisions take place between the primordial cell and the spermatocyst stage. He postulated 13 spermatogonial generations, two of them Type A and eleven Type B. During spermiogenesis the cytoplasm of the nurse cells displays at first a steady increase of lipid droplets which at their greatest development are densely distributed throughout the cell body and then decline. At the end of spermiogenesis few droplets are left. Holstein recognized two functional phases in the life of the epithelial cell, an exocrine phase concomitant with spermatogonial development and a nurse cell phase, possibly endocrine in nature, during spermiogenesis. The morphological and functional change from one phase to the other occurs during the first meiotic prophase. The suggestion that the lipid production of the nurse cells indicates formation of an androsteroid is somewhat supported by the finding of Simpson, Wright & Hunt (1964) who found in the testis of *Squalus acanthias* enzymes which strongly point to the synthesis of testosterone.

Bradyodonti

Spermatogenesis in this class seems never to have been investigated in detail. Stanley (1963) reviewed the small literature and described the testis in *Hydrolagus colliei*. A small proliferative zone, the 'germinal projection', develops on the dorsolateral aspect of the gonad. New germ cells are formed here and are incorporated into ampullae in a manner apparently very similar to selachians. The histology of the ampullae and their migration across the testis to the efferent ductules are also essentially similar.

Teleosts

In the class of the bony fishes, spermatogenesis has been investigated only in the subclass of the teleosts, and even there the coverage has not been broad. A glance at Table 6 shows that representatives of a dozen orders (out of 38) have been explored, and that more than half of the species come from three orders, the Isospondyli, Microcyprini and Percomorphi.

The teleost testis is characterized by the presence of lobules or tubules, the terms are used synonymously in teleosts, within which germ cells develop in cysts (also called follicles; e.g. Stanley *et al.*, 1965) as soon as the spermatogenic process begins.

Usually the cysts are stationary in their position in the tubule and spermiation

TABLE 6. *Key references on teleost spermatogenesis, taxonomically arranged*

Order	Species	Author(s)	Remarks
Isospondyli	*Salmo gairdneri*	Billard, Breton & Jalabert (1970)	Quantitative
	Clupea harengus	Bowers & Holliday (1969)	Histology
	Salvelinus fontinalis	Henderson (1962)	Histology
	Salmo salar	Jones (1940)	Histology
	Notopterus notopterus	Shrivastava (1967)	Histology
	Oncorhynchus nerka	Weisel (1943)	Histology
Haplomi	*Umbra limi*	Foley (1926)	Histology and cytology
	Esox lucius	Lofts & Marshall (1957)	Histochemistry
Ostariophysi	*Conesius plumbeus*	Ahsan (1966*a*, *b*)	Histology, experimental
	Phoxinus laevis	Bullough (1939)	Histology
Apodes	*Anguilla*	D'Ancona (1943)	General reproduction
		Fontaine & Tuzet (1937)	Histology
		Rodolico (1933)	Histology
Microcyprini	*Poecilia reticulata*	Billard (1969*a*, *b*)	Histology, quantitative
	(*Lebistes reticulatus*)	Billard (1970)	Electron microscope
		Billard & Fléchon (1969)	Electron microscope
		Dildine (1936)	Embryology
		Goodrich, Dee, Flynn & Mercer (1934)	Embryology
		Vaupel (1929)	Histology
	Platypoecilus maculatus	Chavin & Gordon (1951)	Embryology
		Wolf (1931)	Embryology
	Gardonus retilus	Clérot (1971)	Electron microscope
	Poecilia sphenops	De Felice & Rasch (1969)	Quantitative
	Gambusia holbrookii	Dulzetto (1933)	Histology
		Geiser (1924)	Histology
	Oryzias latipes	Egami & Hyodo-Taguchi (1967)	Quantitative
	Brachydanio rerio	Ewing (1972)	Histology
	Fundulus heteroclitus	Matthews (1938)	Histology
	Glarydichtys januarius and *G. decemmaculatus*	Phillippi (1908)	General reproduction
	Brachyraphis episcopi	Turner (1938)	General reproduction
Anacanthini	*Gadus merlangus* and *G. esmarkii*	Gokhale (1957)	Histology

TABLE 6. (*cont.*)

Order	Species	Author(s)	Remarks
Percomorphi	*Betta splendens*	Bennington (1936)	Histology
	Blennius	Champy (1921)	Histology
	Lateolabrax japonicus	Hayashi (1971)	Histology
	Gobius paganellus	Stanley, Chieffi & Botte (1965)	Histology and histochemistry
	Micropterus salmoides	Johnston (1951)	Embryology
	Aequidens portaligrensis	Polder (1971)	Histology
	Perca flavescens	Turner (1919)	Histology
	Sargus and *Serranus*	Van Oordt (1929)	Histology
	Cymatogaster aggregata	Wiebe (1968)	General reproduction
	Nandus nandus	Raizada (1975)	Histology, quantitative
Scleroparei	*Cottus bairdii*	Hann (1927)	Histology
	Sebasticus marmoratus	Mizue (1958)	Histology
	Sebastodes paucispinis	Moser (1967)	Histology
	Oligocottus maculosus	Stanley (1969)	Electron microscope
Thoracostei	*Gasterosteus aculeatus*	Craig-Bennet (1931)	General reproduction
	Culaea inconstans	Ruby & McMillan (1975)	Development
Haplodoci	*Opsanus tau*	Hoffman (1963)	General reproduction
Lophiformes	*Lophius piscatorius*	Dodds (1910)	Embryology
Synbranchii	*Monopterus albus*	Chan & Phillips (1967)	General reproduction

takes place into the tubular lumen (Billard *et al.*, 1972), but in at least some *Microcyprini* the cysts migrate from the blind end of the tubule to the efferent duct where rupture occurs (Billard, 1969a).

The arrangement of the lobules shows considerable variation. Brock (1878) recognized two main patterns, one in *Percomorphi*, in which more or less elongate lobules radiate from a peripheral duct, the other in *Microcyprini* in which an anastomosing and branching network of tubules is found. While this categorization may have a certain usefulness, the two types blend into each other and probably represent extremes of a continuous spectrum, nor are they restricted to the orders mentioned. Seasonal reproduction is typical for teleosts as for most vertebrates, and the testes may show extreme seasonal variations in size and histological appearance. For instance, in the perch the average weight of the testis in the mature male varies from 0.12 to 5.88% of body weight (Turner, 1919). On the other hand

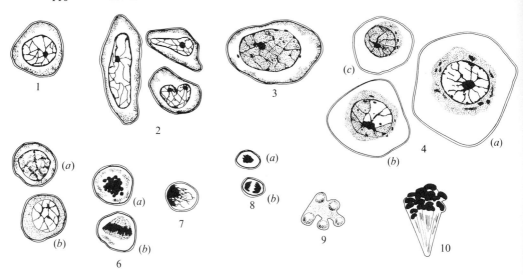

Fig. 25. Drawing of the developing male germ cells of the perch. 1, Germ cells in cord outside of testis before migration. 2, Germ cells during migration to the periphery of the testis. 3, Germ cells at the periphery of the testis just after migration. 4, Transformation of germ cells into spermatogonia: (*a*) early stage; (*b*) intermediate stage; (*c*) late stage. 5, Resting spermatogonia. 6, Dividing spermatogonia at metaphase: (*a*) polar view; (*b*) equatorial view. 7, Nuclei in the bouquet stage of synezesis. 8, Dividing spermatocysts: (*a*) polar view of metaphase; (*b*) equatorial view of anaphase. 9, Transforming spermatids. 10, Group of spermatozoa. (After C. L. Turner (1919). *J. Morph.* **32**, 681–711.)

there are many examples of complete lack of seasonal fluctuations in tropical species (e.g. Turner, 1938; Polder, 1971; Raizada, 1975).

Turner's classical description of the changes observed in the testes of the perch *Perca flavescens* throughout the year, is a favorable introduction to the gross and microscopic anatomy of the male gonad in teleosts because Turner employed a broad comparative point of view. While he concentrated on the specifics of the perch, he examined also the testes of other teleosts, of *Necturus*, two species of turtles, several birds, and the muskrat for comparison.

The two elongate testes of the perch fuse together at their caudal ends and form a single body which communicates with the urogenital sinus through a wide, thin-walled sinus. On the ventral side of each gonad is a deeply embedded connective tissue strand (mediastinum), from which septa radiate to the capsule. Thus the whole organ is subdivided into lobules. The lobules are of irregular shape, joined at their central, pointed apices and diverging and branching toward their broader ends at the periphery. There may be as many as five branchings from a central trunk. In the pike, Turner observed that the branches were much more finely divided and convoluted, 'giving the appearance of seminiferous tubules'.

Piecing together the microscopic observations of consecutive stages, Turner arrived at the following account of spermatogenesis. In mature fish during the spring, germ cells occur in clusters outside the testis within the ventral cord of connective tissue. In subsequent stages they are found in the connective tissue septa and show an irregular, elongate shape characteristic of ameboid motion. The germ cells are moving more and more peripherally and finally accumulate in lens-shaped masses at the blind ends of the lobules, close to the capsule of the testis. Although some mitoses may be seen in these accumulations the main increase of germ cells in the periphery seems to be provided by immigration from outside the testis through the septa. 'It seems that the tendency to migrate ceases only when the cells have definitely come in contact with an obstruction at the end of the tubule.' As soon as the migrating cells have reached the end of their journey they begin to increase in volume and show some cytological changes (Fig. 25). On reaching a maximum size the cells transform into 'spermatogonia'. The proportion of the largest cells never exceeds 5% and transforming cells are rarely seen, an indication that the critical change occurs within a short time. Growth and transformation of cells continue until late November and during the fall each lobule contains a profusion of early germ cells in all growing stages, a few cells in transformation, spermatogonia, and also succeeding stages. Turner found it impossible to determine the exact number of spermatogonial generations but estimated that there were at least five or six. The descendants of each primary spermatogonium become arranged in a cyst incompletely surrounded by connective tissue cells. Cells within a cyst develop synchronously. In each successive generation spermatogonia are smaller than in the preceding one, and spermatocytes are still smaller. There is no growth of the spermatocytes during the whole of the meiotic period (Fig. 25(7) and (8)). The tiny spermatids of each cyst collect into masses which in Turner's words 'resemble parachutes'. These masses become increasingly compact as spermiogenesis proceeds. They are comparable to the structures which have been described in other teleosts as 'spermatozeugmata' (Phillipi, 1908) or erroneously as 'spermatophores' (see Glossary). In the perch the first spermatozoa are seen early in September. They are present in ever increasing numbers in the lumina of the tubules until the following spring when spawning occurs.

Turner's description has provided a general model for spermatogenesis in teleosts. It has served as a standard of comparison for many authors and has been confirmed in general. If we disregard for the moment differences in the modes of seasonal rhythms and in the annual supply of spermatogonia, descriptions are similar for a great number of species which are not necessarily closely related; for instance, *Umbra* (Foley, 1926), *Cottus* (Hann, 1927), *Fundulus* (Matthews, 1938), *Phoxinus* (Bullough, 1939), *Salmo* (Jones, 1940), *Gadus* (Gokhale, 1957), *Clupea* (Bowers & Holliday, 1969), *Salvelinus* (Henderson, 1962), *Sebastodes* (Moser, 1967) and *Aequidens* (Polder, 1971). An exceptional finding has been that in *Cymatogaster* spermatogonia proliferate asynchronously and cysts form in the spermatocytic rather than in the spermatogonial stage (Wiebe, 1968). Despite the uniformity of descriptions there is no clear agreement on several important issues, for instance on the mode of spermatogonial renewal in the cycles of reproductive activity and on the nature of the cysts.

The renewal of spermatogonia has been briefly reviewed by Moser (1967). Turner apparently has remained the only author who saw the source of annual replenishment of spermatogonia in an extratesticular tissue, a mediastinal cord of cells where primordial germ cells lie between fibroblasts. However, he found this stage of development only in one single case and pointed out the inconclusiveness of his finding. Moser observed no extratesticular cords but described germ cells in the fibrous tissue of the testicular capsule between the blind peripheral ends of the lobules. He regarded these as the source of the annual migration of germ cells into the lobules but presented no evidence for their kinetics. Bennington (1936) specifically stated that in the Siamese fighting fish *Betta splendens* 'residual germ cells', much larger than spermatogonia, are present at all times in the walls of lobules, and remain there until the late spermatids are extruded. They then transform into spermatogonia. No evidence was found for any migration of germ cells through the testis. It should be kept in mind that the Siamese fighting fish is a tropical species with a rapid, non-seasonal reproductive cycle. The majority of investigators have come to the conclusion that the spermatogonia arise from germ cells ('stem cells') which have been dormant in the peripheral parts of the testis during preceding spermatogenic cycles.

Ruby & McMillan (1975) have shown by labelling with [^{3}H]thymidine that in the stickleback the spermatogonia in the lobules are derived from clumps of germ cells in the interstitial tissue. These authors did not determine the early kinetics of the germ cells (which they called spermatogonia), but described in detail how these cells enter lobules and how cysts are formed (see below).

Almost all investigators have emphasized that the lobules of the fish testis contain cysts in which single clones of gametes develop, but there has been much uncertainty about the nature of the cyst boundary. Turner considered it an 'incomplete' envelope of connective tissue cells. Many authors avoided the issue of the boundary even when their photographs showed it clearly (e.g. Moser, 1967). Others described that the cysts are surrounded by an apparently complete membrane, without suggesting its nature. Henderson wrote of 'a thin membranous capsule that appears to be continuous with the boundary cell layer'. Stanley *et al.* (1965) found 'a very thin membrane consisting of one or more flattened follicle cells' which extended cytoplasmic processes between the germ cells. Wiebe (1968) considered that the spermatogonia in *Cymatogaster* were enveloped by 'Sertoli cells', but that cysts developed in the spermatocytic stage. The most remarkable statement is that of Polder (1971) that 'the spermatogenic cells lie embedded in the sertolian [*sic!*] cytoplasm, a loose reticular structure without cell boundaries in which the sertolian nuclei can be found everywhere'.

All these results were obtained with the light microscope, and it is quite certain that the resolving power of this instrument only rarely permits clear identification of the somatic cyst cells in teleosts. Fortunately some electron microscopic observations of the subject are now available, and from them we may gain a new insight into the relationship between somatic cells and male gametes in fishes.

Before reviewing the latest results on the composition of cysts we must digress briefly to a question of terminology. Investigators of lower vertebrates have in recent years, usually after only brief consideration, adopted the term "Sertoli cells" for the somatic cells accessory to spermatogenesis which in insects, for instance, usually are named "cyst cells". An extensive discussion of this issue of semantics is presented in Chapter 6.2 of this book (see also Chapter 11). Here I will state only that in view of the variety of cells which we are learning to recognize as essential symbionts of germ cells, it seems prudent to reserve the term "Sertoli cell" for a very specific mammalian type of cell, which Sertoli first described in the human. For teleosts as for amphibians I am using the terms "cyst cell" or "follicle cell" for the somatic component of cyst or follicle, or more generally the non-committal terms "supporting cell" or "auxiliary cell".

Billard (1970) and Billard, Jalabert & Breton (1972) have explored the ultra-structure of cyst cells in guppy, trout, pike, carp, goldfish and roach. They called the cells 'Sertoli cells', although in the summary of the last paper they stated that 'these cells are not genuine Sertoli cells as the spermatogenetic process is not the same as that of higher Vertebrates'. According to these authors the cyst cells during phases of reproductive inactivity are lining the blind ends of the tubules interspersed among single or paired spermatogonia. When spermatogenesis begins the cyst cells surround isogenic groups of spermatogonia, and in this way cysts are formed. During spermatogenesis cyst cells increase in size and number. Initially there are no specific junctions between neighboring cells but there is a multitude of interlocking processes. Later, specific desmosomal junctions appear which in the guppy may cover 50% of the contact surfaces. The tonofilaments splaying out from the desmosomes tie in with bundles of coarser fibrils which form an extensive network close to and parallel with the cell membrane (Plate 14). In all but the earliest spermatogonial cysts the germ cells are only in partial and structurally unspecific contact with the processes of cyst cells which extend into the interior of the cysts. The length of these processes increases from 1 μm in late spermatogonial stages to 20 μm in the later half of spermiogenesis. As the spermatid heads elongate they turn toward the periphery and become inserted between the complex, finger-like processes of cyst cells with which they form a type of "close junction". Billard *et al.* found no ultrastructural evidence for steroidogenesis in the cyst cells, but observed that they become strongly phagocytotic during spermiogenesis. Their role during spermiation has not been clarified although the dependency of this process on fish gonadotrophin and concomitant volume changes in the Sertoli cells has been demonstrated (Yagamazaki & Donaldson, 1968).

Ruby & McMillan (1975) have reported on the behavior of the cyst cells, which they named 'companion cells', in the stickleback. Their results were obtained with light and electron microscope and partly contradict those of Billard *et al.* Having found the predecessors of the spermatogonia in the interstitial (interlobular) tissue (see above). Ruby & McMillan came to the conclusion that the intralobular cyst cells send cytoplasmic processes into the interstitial area where the earliest stages of cyst formation take place. This happens at a time early in the reproductive cycle when the myoid cells, which are a permanent component of the tubular wall, are discontinuously and irregularly distributed. The cyst cell processes penetrate into the interstitium through gaps between the myoid cells. At the spermatocytic stage the myoid cells reorganize, the tubular wall becomes continuous and clearly identifiable, and the cyst cells are lying wholly within the tubules. The authors showed in light micrographs that the cyst cell may at one and the same time begin to surround a spermatogonium peripherally and phagocytize residual spermatozoa centrally. They also observed that spermiogenesis is accompanied by a progressive diminution in the volume of the cyst cells which after spermiation are reduced to a very thin layer of cytoplasm surrounding the nucleus. They are then ready to function again in enveloping another spermatogonial generation. This is in striking contrast to the conclusion of Billard that the cyst cells in the guppy degenerate at the end of spermatogenesis and new cyst cells differentiate at the beginning of the following cycle. Neither results are entirely conclusive and further research is needed. Many of Billard's conclusions have been confirmed in another Poeciliid, *Poecilia latipinna*, by Grier (1975) who, however, was unable to determine the ultimate fate of cyst cells.

Endocrine cells in the teleost testis have been a much discussed subject (for review see Lofts, 1968) since Marshall & Lofts (1956) described 'Leydig cells' in fishes. These cells were identified by the histochemical reactions of their lipid content which undergoes changes parallel to the reproductive cycles seen in birds. Marshall & Lofts described two distinct patterns of distribution of cells in which cyclic changes of lipids occur; (*a*) a typical interstitial gland as found in most higher vertebrates (e.g. in *Gasterosteus*, Craig-Bennet (1931), and (*b*) cells applied to or included in the walls of the lobules (e.g. in *Esox*, Lofts & Marshall (1957)), the 'lobule boundary cells'. In both types, during periods of reproductive inactivity, the lipoidal material is scanty or absent. It increases in the course of active spermatogenesis. In *Gobius paganellus* glandular tissue occurs in small clumps in the interlobular interstitium and also in a large mass, a gland, in the mesorchium (Stanley *et al.*, 1965). The presence of cholesterol-positive lipids and specifically of Δ^5-3β-hydroxysteroid dehydrogenase activity in the glandular cells was regarded as suggestive of androgen production. The bulk of the evidence appears to justify the opinion that teleost interstitial tissue of the type (*a*) of Marshall & Lofts is functionally androgenic, i.e. a Leydig cell homologue, but almost all recent investigators have concluded that the lobule boundary cell is homologous, or at least analogous, to the Sertoli cell of higher vertebrates (Stanley *et al.*, 1965; Billard *et al.*, 1972; Ruby & McMillan, 1975; Grier, 1975). These cells show cyclic sudanophilia in many instances, but at present there is nothing to indicate what specific hormones they elaborate, if any.

A few quantitative data are available regarding the dynamics of the spermatogenic process in teleosts. The number of spermatogonial generations has been counted or estimated repeatedly, usually between five and seven (Table 8, p. 136). Numbers exceptionally high for the whole animal kingdom have been reported for two live-bearing poeciliids of the order Microcyprini: at least 10–12 in *Gambusia* (Geiser, 1924) and 14 in *Poecilia* (Billard, 1969).

Billard's quantitative data are detailed and include counts of germ cells and cyst cells per cyst in progressive stages and measurements of cysts. Cysts increase in diameter as the clones they contain increase in cell numbers. The cyst cells may undergo five successive mitoses beginning at the time of the fifth spermatogonial division. Death is frequent among germ cells, particularly during early spermatogonial stages. The result is an output of spermatozoa of only 65% of the theoretical yield. According to Billard the productivity in the guppy is of the order of magnitude of 150×10^6 spermatozoa per gram of fresh testis per day. For the trout the productivity has been determined as 58×10^9 spermatozoa per gram of testis per year, but only 22% of this number is released. The remainder is absorbed in the testis within the 2–3 months following spawning.

Only very few data are available on the duration of spermatogenesis and some of its stages. Egami & Hyodo-Taguchi (1967) determined in the medaka *Oryzias latipes* the minimum times for the interval from early DNA-synthesis in the leptotene spermatocyte to the early spermatid as being 5 days at 25 °C (12 days at 15 °C) and for the period from early spermatid to spermatozoon, 7 days at 25 °C (8 days at 15 °C). In the guppy, Billard (1968) found that the development from early leptotene to spermatozoon is 14¼ days at 25 °C. The duration of spermatogonial generations and the total duration of spermatogenesis after the first spermatogonial division was calculated from the frequency of cysts occurring at different stages (Billard, 1969). The spermatogonial stages appeared to last 21¾ days, the whole of spermatogenesis 36 days at 25 °C. In *Poecilia shenops* the duration from leptotene to mature spermatozoa is at least 21 days, as determined by [³H]thymidine labeling (De Felice & Rasch, 1969).

Amphibia

While the utilitarian value of teleosts as food and game has promoted much interest in their reproduction, amphibians have received less consistent attention. On the other hand, their easy maintenance in the laboratory and the relatively large size of their germ cells have made amphibians attractive objects from the very beginning of cytological and genetic investigations, and recently they have proved to be excellent experimental animals for research in the endocrinology of reproduction. A number of important contributions to basic biology have been achieved in the course of investigations of spermatogenesis in amphibians. For instance, Vallette St George (1876) described the various stages of spermatogenesis in five amphibian species from observations on testicular tissues, fresh or fixed in osmium tetroxide, and laid the groundwork for modern terminology in the field, coining the words 'spermatocyst', 'spermatogonium' and 'spermatocyte'; Flemming (1887) used the spermatocytes of *Salamandra maculosa* in his meticulous study of the meiotic divisions, and Witschi (1914) first recognized in *Rana temporaria* that the male sex in frogs is digametic. Yet the literature lacks any single article which might serve as a classic comprehensive description of amphibian testis and spermatogenesis.

In amphibians, as in all vertebrates, spermatogenesis occurs in cycles. In the majority of tropical and in a few non-tropical species the cycles are continuous and non-seasonal, but species with seasonal cycles of the temperate zones, particularly anurans, have been much more thoroughly investigated and are better known. In the majority of these species the spermatogenic cycle is of the "post-nuptial" type, i.e. spermatogenesis begins immediately after spawning and is completed in a few months so that the tubules contain spermatozoa in an advanced state of maturity for long periods before spawning. The patterns of cyclical activity and their variations and distribution have been comprehensively reviewed by van Oordt (1960). Details of the control of spermatogenesis are beyond the scope of this chapter, although some features will be touched upon below (see spermiation), and will be discussed more extensively in Chapter 11. However, in studying the morphology and the kinetics of spermatogenesis in amphibians and in all vertebrates it must be kept in mind that testicular function depends in part on the secretory activity of the pituitary gland as well as on androgenic secretions usually produced within the testis. (See review by Lofts, 1968.)

The morphology of the amphibian testis is similar to that of teleosts. Spherical or tubular compartments are radially arranged and open separately into a more or less peripheral duct system which runs longitudinally toward the efferent duct at the caudal end of the gonad. The compartments are usually called "ampullae" in urodeles and "seminiferous tubules" or, rarely, "lobules" in anurans. Within them the germ cells develop in cysts. In the intertubular, interstitial tissue "Leydig cells" are found which display distinct lipid cycles and have been regarded for a long time as secretors of androgen (Humphrey, 1921; Lofts & Boswell, 1960; Lofts, 1964).

The testes of amphibians are usually of simple ovoid or elongate shape like those in fishes, but in some urodele species, e.g. in *Salamandra maculosa* (Meves, 1896), they are distinctly lobed, i.e. there is a tandem arrangement of three to five successive enlargements which appears as though multiple testes were lined up along the gonadal axis. Humphrey (1922) has shown that this mode of testis formation is due to a special modification of the process of spermatogenesis. In all urodeles there is a so-called "spermatogenic wave" along the length of the testis which results in the most progressive developmental stages being located in the caudal, the youngest stages in the cranial region. The speed of the wave, however, differs greatly in various species. In *Plethodon*, for instance, the regeneration of lobules after spermiation proceeds rapidly, and spermatogenic activity is almost continuous. Year after year the spermatogenic wave extends over a somewhat longer distance, and the testis simply lengthens. In *Desmognathus* and *Diemyctylus*, however, spermiation is not followed immediately by resumption of spermatogenesis. In fact the next period of spermiation in the area may not occur for another 3 years. In the meantime the spermatogonia in the adjacent, more cephalic, region produce an actively spermatogenic enlargement of the testis in the course of one year, and another one, still more cephalic, in the course of the second year. The narrow necks between the enlargements or lobes contain stem cells capable of future spermatogenic development and are not "sterile". The multiple testes are clearly the result of the spermatogenic pattern of the species. Modifications in the pattern cause differences in the number and size of lobules and consequently of lobes, even in animals of the same species and age. In addition, it has been shown that peculiar shapes of the testis which occur in several species of urodeles are due to local degenerations of germ cells which result from temporary unfavorable conditions (Humphrey, 1925).

The testis of *Proteus anguineus* is elongate and unlobed. It is a particularly favorable object for study and Stieve (1920) has given a brief description of it in which he emphasized the very rich blood and lymph vascular network surrounding the ampullae. According to Stieve, the primary spermatogonia are unusually large (25–35 μm). They divide occasionally and by these means the population remains at a steady level although degenerations occur. At the beginning of the reproductive season each spermatogonium becomes enveloped by two to four cyst cells (called 'follicle cells' by Stieve). Gonial multiplication and cyst formation begin at the cranial end of the testis and rapidly progress over its whole length. When the spermatocytic stage is reached cysts contain approximately 64 or 128 cells and are surrounded by four to six cyst cells. At this stage blood and lymph spaces around the ampullae are distended. Stages of long duration appear quite synchronous in each ampulla, but the synchrony in individual cysts is much more rigorous. No interkinetic phase was observed during the cell divisions of spermatogonia, which diminish rapidly in volume. In early stages this diminution occurs mainly at the expense of the cytoplasm but later the nuclei also became smaller. The diameter of cells of the last spermatogonial generation is approximately half that of primary spermatogonia. The spermatocytes I grow slowly until they have gained the same volume as the primary spermatogonia. Two types of spermatid cysts were found, those which contained on the average 238 germ cells (233–243) and those with an average of 484 (472–504). This indicates that some clones may divide six, others seven, times and further suggests that between 6 and 7% of the cells degenerate in the process. In the spermatocytic stage of *Salamandra*, *Amblystoma* and *Triton*, Gurwitsch (1911) observed only cysts of the large type, which indicated that in these species seven spermatogonial divisions are the rule.

The primary gonia of amphibia show peculiar nuclear characteristics which may not be unique among animals but, perhaps because of the large cell size, are particularly conspicuous in this group. The nuclei are unusually large and poly-

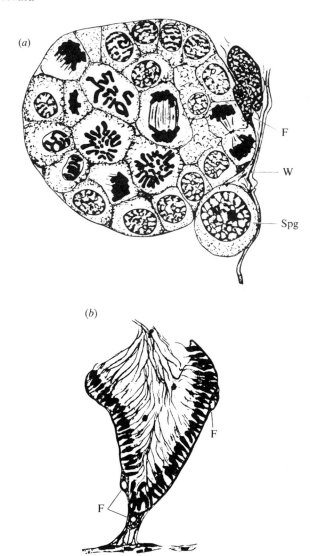

(a)

F

W

Spg

(b)

F

F

Fig. 26. Spermatocysts of *Rana temporaria*. (a) A cyst of spermatocytes in somewhat different stages of development. F, cyst or "follicle" cell; Spg., "free" spermatogonium; W, wall of tubule. (b) A bursting cyst with immature spermatids. The cyst is attached to the wall of the tubule below. Cyst cells are characteristically vacuolated. (After E. Witschi (1914). *Arch. mikrosk. Anat.* **86**, 1–50.)

morphic. The polymorphism has been observed in many species of anurans and urodeles (Poska-Teiss, 1933), and Sentein & Temple (1971) have published a few electron micrographs of the phenomenon in *Triturus*. According to Poska-Teiss the interkinetic nuclei contain chromosomal vesicles which arise in telophase. The number of vesicles is haploid. Mitotic details are in general quite different from ordinary somatic mitoses. Usually the development of chromosomal vesicles, the nuclear polymorphism, and the size of the nuclei diminish conspicuously in successive generations of secondary gonia, but this is not pronounced in *Bufo vulgaris*.

A brief description of spermatogenesis in the anuran *Rana temporaria*, with excellent photographs of tubules in various stages of spermatogenesis, was given by Witschi (1924). For our purposes the most interesting details in his account, which is largely devoted to the behavior of chromosomes, pertain to the process of spermiation. In most animals it is difficult to observe the release of spermatozoa from the somatic cells, but in Amphibia the cyst cells and the germ cells are so large that they can be resolved easily by the light microscope. In the frog, according to Witschi, the cyst wall ruptures as soon as the flagella of the spermatids begin to grow, and opens into the lumen of the tubule (Fig. 26). At the peripheral pole of the cyst the cyst cells aggregate and then transform into large, pillar-like cells which Witschi called 'Sertolische Stützzellen'. There are on the average 12 of these in every cyst and to each attaches a bundle of 60–150 spermatids. Each bundle, with the heads of the spermatids pointing toward the periphery of the cyst, is deeply embedded in a cup-like depression of the supporting cell. The remarkable transformation of the cyst cells into carriers of spermatids took on much greater significance when De Robertis, Burgos & Breyter (1946) suggested that these cells are the target of gonadotrophic stimulation and, therefore, mediate at least part of the seasonal cycle of spermatogenic activity. It was shown that the cyst cells in the spermatid phase (here called supporting cells) develop large cytoplasmic vacuoles which rupture and release fluid and spermatozoa into the lumen of the seminiferous tubule. Details of the cycle of the cyst cells and its correlation with the spermatogenic process will be discussed in the following.

Witschi (1924) described that in *Rana* during the winter months preceding the spawning period ('rest period') sperm bundles in supporting cells line the open cysts which at this stage constitute the wall of the seminiferous tubules. At the periphery, primary spermatogonia surrounded by cyst cells are occasionally seen in preparation for the next reproductive period which begins in summer. The supporting cells were explored with the electron microscope by Brökelmann (1964). Toward the end of the rest period these cells are connected by multiple desmosomes. Their nuclei are large, lobed and contain a prominent nucleolus as well as accumulations of glycogen. The cytoplasm is somewhat vacuolated and shows osmiophilic droplets and large masses of glycogen granules. The mitochondrial cristae are usually tubular. In the spawning period the cells appear even richer in inclusions. Vacuoles accumulate largely around the heads of the spermatids and

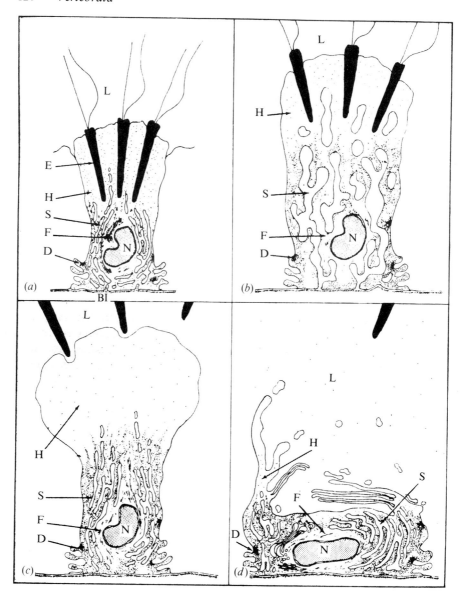

Fig. 27. S, endoplasmic reticulum; H, apical cytoplasm; F, bundles of fibrils; E, apical recesses containing head of spermatid; BI, basal infoldings; D, desmosomes; N, nucleus; L, lumen (of cyst in (*a*), of tubule in (*c*) and (*d*)). Further explanation in text. (After M. H. Burgos & R. Vitale-Calpe (1967). *Am. J. Anat.* **120**, 227–52.)

often show tubular evaginations. After spermiation the supporting cells contain many phagosomes filled with membranes and vacuoles, the remains of spermatid cytoplasm. Glycogen disappears from the cells in the process of spermiation. It is released with the spermatozoa (van Dongen, Ballieux & Geursen, 1960).

The morphology of the spermiation process has been investigated more thoroughly by Burgos & Vitale-Calpe (1967) who also briefly and fairly completely reviewed previous research on this subject. They demonstrated that bovine luteinizing hormone (LH) stimulates spermiation in toads, and that the supporting cell cytoplasm is a target of this hormone, which causes an increase in water and sodium content correlated with morphological changes in the cells. Fig. 27 indicates the essential ultrastructural features of the process. Fig. 27(*a*) shows the stage of the swelling of the endoplasmic reticulum which pushes the spermatids away from the nuclear zone. Fig. 27(*b*) indicates the swelling of the apical cytoplasm (H) which occurs concomitant with a collapse of the vacuolar network and ends in the detachment of the apical portion of the cell (Fig. 27*c*). It appears that the integrity of the basal, nucleated part of the supporting cells is preserved by the apposition of membrane systems. The reduced supporting cells subsequently are transforming into cyst cells which apply themselves to primary spermatogonia. This transformation and the development of the cyst cells throughout the spermatogonial and spermatocytic stages of spermatogenesis need further exploration, but it is clear from the available observations that a series of complex interactions between cyst cells and germ cells occurs at successive stages of development.

Quantitative data on the kinetics of spermatogenesis in Amphibia are rare. Most are contained in the comprehensive thesis of van Oordt (1956) which unfortunately has not been available to me. Bustos & Cubillos (1967) determined the duration of the cell cycle in primary and secondary spermatogonia of *Bufo spinolosus*. Their radioautographic study permitted them to conclude that spermatocytic development from early leptotene to metaphase I takes at least 3 days. In the same animal it was shown that primary spermatogonia exhibit diurnal variation of mitotic activity, but secondary gonia do not (Bustos-Obregón & Alliende, 1973). Spermatogonial activity is decreased but not suppressed by low temperature (4 °C), but does not seem to be influenced by high temperature (37 °C) in comparison to a standard temperature of 18 °C (Bustos-Obregón, Alliende & Schmiede, 1973).

While the spermatogenic process in amphibians is in general very similar to that in teleosts, some phase of it, for instance spermiation, are better known and the endocrinology has been more thoroughly clarified, even if much remains to be done. The most conspicuous difference is the relatively early rupture of the cyst which essentially transforms the cyst wall into part of the tubular wall and exposes the developing spermatids to the tubular lumen. By contrast, the cysts in teleosts usually remain intact until spermiation.

Amniota: Reptilia, Aves and Mammalia

The amniotes, which have many features in common – for instance homoiothermy, a complex genital apparatus, an endocrine control of gonadotropins and androgens, and internal fertilization – show such uniformity in their pattern of spermatogenesis that they are best discussed together. In all of them the gametes develop in seminiferous tubules, not in closed cysts, but in several layers which represent successive generations. These evolve cyclically in concert and in symbiosis with somatic ''Sertoli cells'' which are individually in contact with germ cells of all stages.

The presentation will follow a different plan from that used in previous chapters. In reptiles and birds research has centered almost exclusively on the environmental and endocrine control of seasonal cycles of reproduction. The configuration and the kinetics of the cell population of the seminiferous tubules have been investigated only to the point where their essential similarity with mammals became assured. The mammalian testis, on the other hand, has been explored by a greater number of investigators and in far more detail than the male gonads of all other animals together. The vigor of research in this field has increased manyfold in the last two decades, partly because of the development of new instrumentation and refined methodology and partly because of a rising tide of social and political interest in human reproduction. This recent progress in mammalian research has been reviewed almost to excess. While in 1962 I looked back to the last previous summaries written 60 years ago, today I can refer to at least 10 major reviews which have appeared since 1967. It is, therefore, superfluous to add another review at this time. Instead, the following pages will provide a guide to the intricacies and problems of spermatogenesis in amniotes *without* detailed descriptions. Reference will be made only to the most pertinent and leading literature, and particularly to reviews in the field. Certain topics of spermatogonial renewal, cellular kinetics and the role of the Sertoli cells will be discussed in more detail in Chapters 9 and 11.

The testes of amniotes are compact, more or less ovoid organs surrounded by a heavy investment, the tunica albuginea, which often contains smooth muscle in specific arrangements (Davis, Langford & Kirby, 1970). In animals with seasonal spermatogenesis, particularly in birds, the volume of the testis varies remarkably and the structure of the tunica must be correspondingly adaptable. In the chaffinch, for instance, the testis increases in volume approximately 1000-fold between January and May (Disselhorst, 1908).

The pattern of the capillary network around the seminiferous tubules appears to be similar in all amniotes, but the distribution of the major vessels, particularly the arteries, varies greatly. In mammals with scrotal testes the testicular artery takes a prolonged course within or immediately underneath the tunica, and in some species circles the testis several times before its branches penetrate to the interior (Setchell, 1970; Gunn & Gould, 1975). The lymphatic spaces in the interstitium are extensive and have been investigated ultrastructurally in several mammals (Fawcett, Heidger & Leak, 1969; Fawcett, Neaves & Flores, 1973). Extent and pattern of the lymphatics are certainly related to the function of the androgenic cells of Leydig which have been shown in the rat to have free access to the lymph (Clark, 1976).

The seminiferous tubules are long and display varying degrees of coiling. Essentially each tubule represents a loop with two limbs which usually enter separately into a network of narrow, anastomosing passageways, the rete testis, which in turn connects with the efferent ducts. The number of tubules varies considerably, e.g. 32 in the rat (Roosen-Runge, 1961) and many hundreds in the human. The length of the tubules is even more variable, but the average diameter stays within the range of 180–300 μm. In the majority of species the tubules are unbranched, but they may form a complex anastomosing system, e.g. in man (Johnson, 1934; Hedinger & Weber, 1973).

The seminiferous tubules are enveloped by a complex "limiting membrane" which varies in thickness and composition in different species but always appears to contain myoid cells. The membrane has been most thoroughly investigated in the rat (Ross, 1967) and in man (Bustos-Obregón & Holstein, 1973; Bustos-Obregón, 1974). Unsicker & Burnstock (1975) have shown that in reptiles the myoid cells of the peritubular tissue differentiate in the spring before the reproductive season and that in the fall most of them are 'replaced' by fibroblast-like elements. This paper also has a succinct review of the literature on the subject.

The immature or the seasonally inactive seminiferous tubules consist of a population of gonocytes or primary spermatogonia of types that must be identified specifically in every case (for the rat, see Hilscher *et al.*, 1974) and of somatic cells, the young Sertoli cells, which at this stage have been called 'undifferentiated epithelial cells' (Stieve, 1930). Active spermatogenesis begins with spermatogonial multiplication. Each time when a synchronous clone or an assemblage of clones of spermatogonia has reached the spermatocytic stage, the cells move inward and form a new layer. This occurs at specific and regular intervals. The thickness of the seminiferous tissue increases until four or five generations have evolved. When the first generation is released into the lumen of the tubule, a new one takes its place at the periphery. The Sertoli cells extend radially through the whole width of the tubular wall. They form pillars of cytoplasm with elaborate extensions which entirely enfold spermatocytes and spermatids, but do not interpose themselves between the spermatogonia and the basement membrane (Fig. 28). The essential uniformity of this pattern of spermatogenesis which represents a "cycle of the spermatogenic tissue", is illustrated in Figs. 29(*a*) to (*h*) in comparable stages of the duck, the rat and man.

Accurate identification of stages of the cycle is accomplished by a method established by Leblond and Clermont (1952*a*, *b*) and Clermont & Perey (1957). The stages of acrosomal development, particularly after staining with PAS, are easily identified with the light microscope and not only define the specific stage

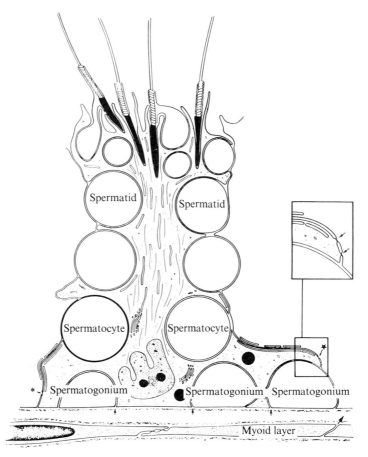

Fig. 28. Diagram of Sertoli cell and its microenvironment. Note relationships with germ cells. The Sertoli cells and their right junctions (*) delimit a *basal* compartment in the seminiferous epithelium, containing spermatogonia and preleptotene spermatocytes, and an *adluminal* compartment, containing spermatocytes and spermatids. (After M. Dym & D. W. Fawcett (1970). *Biol. Reprod.* **3**, 308–26.)

of the spermatid, but also mark unmistakably stages of spermatocytes and spermatogonia, because all generations of germ cells are in step within the cycle. The underlying features of this method of staging as seen with the electron microscope are illustrated in Figs. 30 to 33. At present the cycles of only a few mammals have been explored thoroughly, and no reptilian and avian patterns have been studied in detail. A modified view of spermatogonial proliferation and renewal in the rat has been proposed by Moens & Hugenholtz (1975). The view is compatible with concepts of Clermont (1972) and Huckins (1971*a*, *b* and *c*), but contains some essential differences. The established concept, as represented diagrammatically

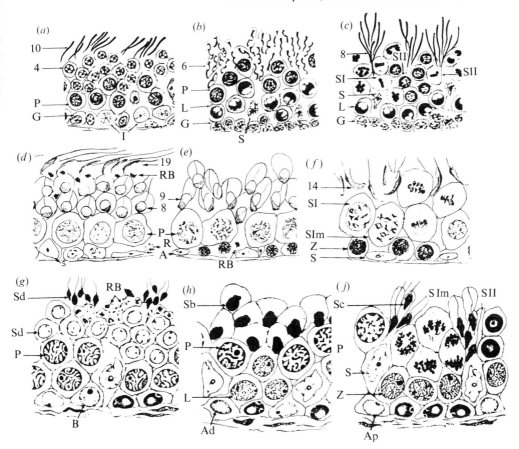

Fig. 29. Semi-diagrammatic representations of three very similar stages of the spermatogenic cycle: (a)–(c) in the duck (Clermont, 1958); (d)–(f) in the rat (Leblond & Clermont, 1952); and (g)–(j) in man (Clermont, 1963). Numbers in the duck and the rat refer to spermatids of different stages. They are not equivalent in the two animals. G, spermatogonia; Ad, Ap and B, spermatogonia of different stages; SI and SIm, primary spermatocytes; L, leptotene; Z, zygotene; P, pachytene; SII and SIIm, secondary spermatocytes; S, Sertoli cell; Sb, Sc, Sd spermatids of different stages; RB, residual bodies. (After Y. Clermont (1958). *Arch. Anat. microsc.* **47**, 47–66; C. P. Leblond & Y. Clermont (1952). *Ann. N.Y. Acad. Sci.* **55**, 548–84; and Y. Clermont (1963). *Am. J. Anat.* **112**, 35–45.)

in Fig. 34, is that a certain proportion of a base population of spermatogonia (As or Apr and Aal) periodically enter the spermatogenic pathway of differentiation. They then pass through all spermatogonial, meiotic and spermiogenic stages. The data of Moens & Hugenholtz, derived from electron micrographs of serial sections, appear to indicate that there are quantitative and qualitative differences in the spermatogonial population at stages which are identical in terms of spermatocyte

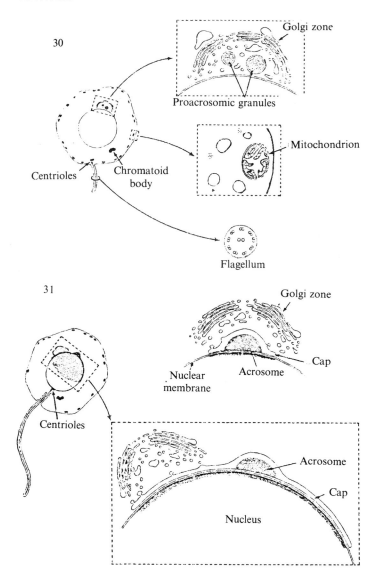

30

Golgi zone

Proacrosomic granules

Mitochondrion

Centrioles

Chromatoid
body

Flagellum

31

Golgi zone

Cap

Nuclear
membrane

Acrosome

Centrioles

Acrosome

Cap

Nucleus

and spermatid morphology. The authors' interpretation of the data is that for a given area in a seminiferous tubule there is a 'periodic build-up of spermatogonia' which then produce several successive quanta of later stages. When the spermatogonia are depleted the process is repeated.

While this modification of the model of spermatogonial renewal may seem to be slight, the authors also propose that non-random degenerations of cells do not occur in normal spermatogenesis of the rat, a basic contradiction of the generally

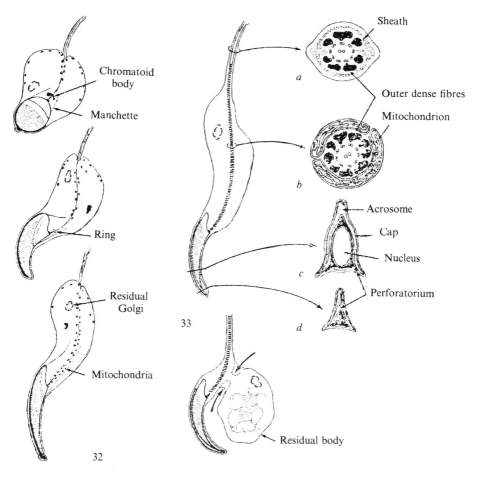

Figs. 30–33. Semi-diagrammatic representations of cytological differentiation of rat spermatids at various steps of the process. Numbers indicate the steps as shown also in Fig. 34. 30 represents the 'Golgi phase'; 31 the 'cap phase'; 32 the 'acrosome phase'; and 33 the 'maturation phase'. (After Y. Clermont (1967). *Arch. Anat. microsc.* **56** (Suppl.), 17–60.)

accepted view. The fact that the numbers of cells produced in the cycle progressively fall behind the number calculated from the number of cell divisions, Moens & Hugenholtz explain by differences in the cells within a syncytial clone. 'Within a dividing syncytium a few cells do not divide while they advance developmentally with the syncytium as a whole.' While the data on differential cell behavior within a clone are indeed meaningful, the denial of the occurrence of systematic degenerations is not conclusive on the basis of the exploration of 13 samples of seminiferous tissue (about 0.05 mm³ each). On the contrary the quantitative and

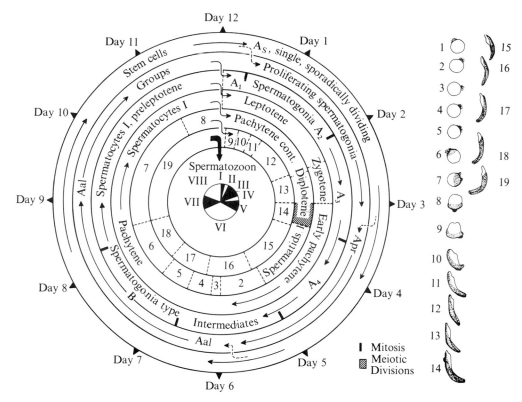

Fig. 34. Composite diagram of the cycle of the seminiferous epithelium in the rat. The stages of spermiogenesis (Leblond & Clermont, 1952) are depicted on the right. The circle in the center indicates spermatogenic stages according to Roosen-Runge & Giesel (1950) and Ortavant (1958). The outer two circles are constructed according to the views of Huckins (1971*a*, *b*). The events of spermatogenesis may be followed from stem cell to spermiation by starting at the top and moving clockwise in a spiral inward. Any radius indicates stages which may occur together in a constellation within a given segment of a tubule, as for instance in the three stages of spermatogenesis in Fig. 29(*c*), (*d*) and (*e*).

qualitative evidence for such degenerations is still very strong, as presented in Chapter 10, pp. 146–7.

While individual gametes in their development move from the periphery of the tubule to the center, there also exists in many species a more or less pronounced progression of successive stages of the cycle along the axis of the tubule. This is the *spermatogenic wave*. It is the result of synchronous development of groups of germ cells and the progressive spread of this development to adjacent areas of the tubule. When groups are large and progression is regular, as in the rat, a fairly regular and conspicuous wave results. When groups are very small, as in man, a wave may not be manifest. In most mammals intermediary conditions are seen. The wave has been most thoroughly studied and quantitated in the rat (Perey, Clermont & Leblond, 1961).

Reviews on spermatogenesis in reptiles and birds are confined to seasonal cycles and their correlation with exogenous and endocrine factors (Lofts, 1968, 1969; Ozon, 1972; Farner, 1975). Some detailed studies on reptiles offer a lead to the literature, e.g. Herland (1933), Risley (1938), Cavazos (1951), Wilhoft & Reiter (1965). The classical account of Dalcq (1921) on spermatogenesis of the slow-worm *Anguis fragilis* contains a detailed cytological description and a still-valid discussion of basic issues (see also Chapter 10). Of all birds only the duck has been investigated in some detail comparable to mammalian studies (Schöneberg, 1913; Clermont, 1958).

Table 7 lists the mammalian species in which the cycle of the seminiferous tissues has been relatively thoroughly explored, the main investigators and their specific topics. This survey obviously omits the multitude of recent contributions to the cytomorphology and cytophysiology of germinal and Sertoli cells. However, the following annotated chronological list of review articles may indicate the extent of the investigative effort and provide access to the bulk of the literature.

Roosen-Runge (1962) 33 pp. Covers the background necessary for an understanding of terminology and results of quantitative and high resolution investigations which began around 1950. Contains the first "spermatogenic clocks" of rat, mouse and ram.

Clermont (1967) 54 pp., in French. Concentrates on quantitative kinetics. Contains an excellent brief description of mammalian spermiogenesis (Figs. 30–33).

Thibault (1969) 32 pp., in French. Treats many aspects of mammalian spermatogenesis but is not broadly comparative. There is a very brief review of the tubule in rat and mouse, a section on ultrastructure of gametes and the Sertoli cell (28 pp.) and two unique composite diagrams of the cycle in ram and man. The phenomenon of the spermatogenic wave and the renewal of spermatogonia are discussed. There is also a section on maturation and transport of spermatozoa in the male tract.

Courot, Hocherau-de Reviers & Ortavant (1970) 92 pp. The cells of the tubule, their morphology and histochemistry are described. Kinetics in several species are presented in tables and diagrams with special emphasis on the duration of stages and the methods used to determine the duration. The review is truly comparative and contains some material not included elsewhere.

Setchell (1970) 137 pp. An extensive and unique review of testicular blood supply, lymphatic drainage and the secretion of fluid. It contains the background for an understanding of the environment of the developing germ cells and for the physiology of the barrier between blood and the spermatogenic process. This review has recently been updated and supplemented by a briefer article on the blood–testis barrier (Setchell & Waites, 1975).

Clermont (1972) 38 pp. A review of the cycle, and of spermatogonial renewal

TABLE 7. *Key references on mammalian spermatogenesis*

Species	Author(s)	Subject
Boar	Swierstra (1968)	Kinetics
Bull	Kramer (1960)	Descriptive, quantitative, stages
	Hochereau-de Reviers (1970)	Gonial renewal
	Brendston & Desjardin (1974)	Kinetics
Coyote	Kenelly (1972)	Kinetics duration
Elephant	Johnson & Buss (1967)	Descriptive
Guinea-pig	Cleland (1951)	Descriptive, cycle
	Clermont (1960)	Kinetics
Hamster	Clermont (1954)	Gonial renewal
	De Rooij (1968)	Kinetics
	Clermont & Trotter (1969)	Kinetics
Macaque	Clermont & Antar (1973)	Kinetics
Man	Clermont (1963)	Cycle
	Heller & Clermont (1964)	Kinetics
	De Kretser (1969)	Electron micrographs, spermiogenesis
	Holstein & Wartenberg (1970)	Electron micrographs, spermatogenesis
	Rowley, Berlin & Heller (1971)	Electron micrographs, gonia
	Rowley & Heller (1971)	Kinetics
Monkey species	Barr (1973)	Kinetics
Mouse	Oakberg (1956)	Descriptive kinetics
	Monesi (1962)	Cell cycles
	Monesi (1965)	RNA and proteins
	Oakberg (1971)	Gonial renewal
	De Rooij (1973)	Gonial renewal
Rabbit	Swierstra & Foote (1963)	Cycle
Ram	Ortavant (1958)	Descriptive, kinetics
	Setchell (1967)	Fluid barrier
Rat	Roosen-Runge & Giesel (1950)	Stages, quantitative
	Leblond & Clermont (1952*a, b*)	Stages, cycle
	Clermont (1962)	Kinetics
	Clermont & Bustos-Obregón (1968)	Gonial renewal
	Hilscher & Makoski (1968)	Fetal and adult, quantitative
	Huckins (1971*a, b, c*)	Gonial renewal, cycle
	Dym & Fawcett (1970)	Sertoli, electron micrographs
Stallion	Swierstra, Gebauer & Pickett (1974)	Kinetics

in rat, mouse, ram, bull and monkey with many diagrams of various models of the process.

Dalcq (1973) 834 pp., in French, 57 pp. of references. This posthumous tome contains a huge amount of information on the normal cytomorphology of testis and spermatogenesis. The development of the gonad is concisely reviewed in a comparative way. There are large chapters on Leydig and Sertoli cells, spermatogonia, spermatocytes, spermatids and spermatozoa and stimulating general discussions. Ultrastructural research is generally underemphasized, although definitely represented in the chapters on spermiogenesis. The review is uniquely strong in cytochemical considerations and methodology. Many interesting ideas and perspectives are distributed almost casually throughout the book, the wisdom of a great scientist at the end of five decades of work in the field. (Caution: the references are not always accurate.)

Fritz (1973) 44 pp. A review of certain aspects of the biochemistry of spermatogenesis. It contains a section on techniques for obtaining relatively pure populations of various cell types from the testis, and some examples of the application of such procedures to investigations of phases of spermatogenesis. Another section deals with enzyme changes in spermatocytes and spermatids.

Steinberger & Steinberger (1974) 20 pp. and *Steinberger & Steinberger* (1975) 20 pp. The first of these reviews deals with hormonal control of testicular function. The data are largely derived from studies on the rat and on man and are not comparative in detail, but rather a summary of current general concepts. The second review supplements the first with respect to morphology and kinetics. Some results of the culturing of testicular tissues are mentioned. Together the reviews bring a functional view to the structural and quantitative data.

Fawcett (1975*a*) 34 pp. A succinct, up-to-date review of the Sertoli cell, essentially from the morphological point of view, but with consideration of all its possible functions.

Fawcett (1975*b*) 41 pp. A concise, superbly illustrated review of the morphology of the mammalian spermatozoon.

Bustos-Obregón, Courot, Fléchon, Hochereau-de Reviers & Holstein (1975) 21 pp. Restricted largely to ultrastructural morphology with special emphasis on spermatogenesis in man. An additional aim is the standardization of the terminology on an international level.

With this satellite view of the intense investigative work on mammals ends the phylum-by-phylum treatment of the process of spermatogenesis in animals. It is now abundantly clear that only in mammals have modern instruments and techniques been applied somewhat systematically. The following chapters of comparative considerations should be read with this perspective in mind. On the one hand, the relative paucity of information on all groups other than mammals forces us to great caution with regard to generalizations. On the other hand, the wealth of ideas springing from the store of mammalian data concerning, for example, the barrier between blood and germinal tissues or the complexity of function of Sertoli cells, enables us to see many possible analogies and lines of pursuit in other animals, and has already enriched our general concepts of spermatogenesis.

Plate 1. The beginning of research on the morphology of spermatogenesis. (From Rudolph Wagner (1839). *Erläuterungstafeln zur Physiologie und Entwicklungsgeschichte.* Leipzig: Voss.) Wagner's explanations (in translation from Latin and German, slightly abbreviated): Fig. III spermatozoa of several mammals from the vas deferens. *a* represents usually the broad side, *b* the narrow side view: 1, red monkey, *Cercopithecus ruber.* 2, horse-shoe nose bat, *Rhinolophus ferrum equinum.* 3, mole *Talpa europaea.* 4, dog, (Pomeranian); in *a* one sees a round spot. 5, rabbit, *Lepus cuniculus*; the shadow got a little too strong. 6, house-mouse *Mus musculus*; *b* from the back; one sees the root of the tail a little higher up. 7, rat, *Mus rattus*; the body in *a* and a* in profile view. 8, field-mouse, *Hypudeus arralis.* Fig. VI. Rabbit semen. A, from the *vas deferens. a* and *b*, Two pale, flat, finely granulated bodies which I consider epithelial flakes ('Epithelblättchen'). *c*, Two darkly bordered corpuscles throwing deeper shadows, which I consider genuine particles of the semen, 'granula seminis' [they look like spermatids, E.R.R.]. *d*, Blood corpuscles for comparison, one in profile. *e*, A spermatozoon. B, Semen from the testis. *a* and *b*, Two corpuscles with very dark borders which I regard as fat droplets. *c*, A group of much smaller, similar corpuscles, probably also fat particles. *d* and *e*, Two granular pale bodies, probably seminal bodies ('Samenkörnchen') or, perhaps, epithelial flakes. *f*, A similar one, only darker. *g* and *h*, Two developmental spheres or cysts [probably artifacts involving spermatids and possibly Sertoli cells, E.R.R.]

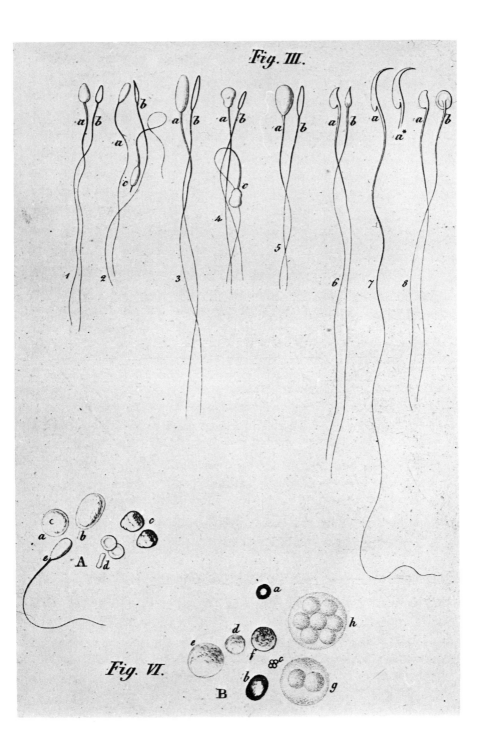

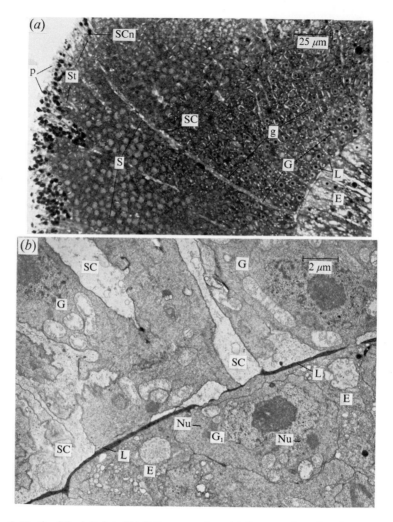

Plate 2. Testis of the jellyfish *Phialidium gregarium*. (*a*) Cross-section of testis; epon-section 1 μm. An epithelial supporting cell (SC) is seen to run from the surface of the testis at left to the mesogloeal lamella (L) at right (approx. 160 μm). Its nucleus (SCn) is near the surface. G, zone of primary spermatogonia; g, zone of secondary spermatogonia; S, zone of spermatocytes; St, zone of spermatids; p, pigment granules in surface part of supporting cells; E, gastrodermal epithelium. (*b*) Electron micrograph. Processes of supporting cells (SC) abut against dense mesogloeal lamella (L). The lamella is of varying thickness and occasionally discontinues. Gastrodermis below (E) contains a germ cell, probably a "stem cell" (G_1) in which "nuage" is visible (Nu). G, spermatogonia within gonad. (From E. C. Roosen-Runge & D. Szollosi (1965). On biology and structure of the testis of *Phialidium* Leuckhart (Leptomedusae). *Z. Zellforsch. mikrosk. Anat.* **68**, 597–610.)

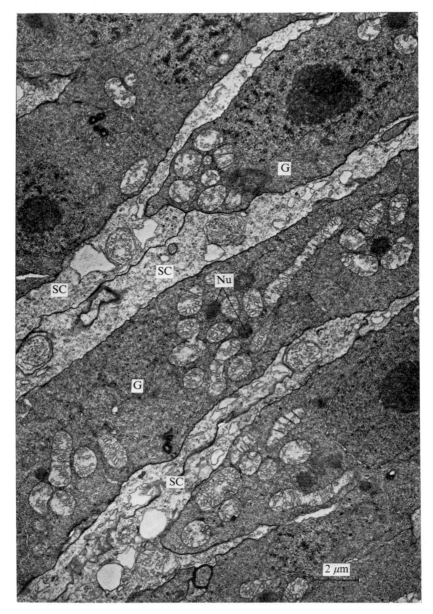

Plate 3. Electron micrograph of the spermatogonial zone (cross-section of testis) of the jellyfish *Phialidium gregarium*. Supporting cells (SC) are remarkably closely applied to each other and to spermatogonia (G) which they do not completely envelop. The gonia have very large nucleoli and contain nuage (Nu). (From E. C. Roosen-Runge & D. Szollosi, unpublished data.)

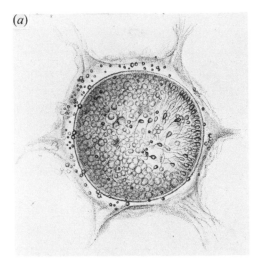

(a)

Plate 4. (a) Drawing of spermatocyst of the freshwater sponge *Spongilla lacustris* in fresh condition. The heads of spermatids are applied to the capsule. The remainder of the contents of the cyst is not identified. (From N. Lieberkühn (1856). Beiträge zur Entwicklungsgeschichte der Spongillen. Addenda. *Müller's Arch. Anat. Physiol.* 496–514.)

Plate 4. (b), (c) and (d) Atypical, apyrene "carrier" spermatozoon of the snail *Janthina* (see Chapter 10, p. 150 for description). The scale represents 200 μm. (b) Before the attachment of typical spermatozoa: 1, undulating membrane, 'Treibplatte'; 2, mass of yolk; 3, connecting piece; 4, tail, 'Ansatzstück'. (c) After the attachment of typical spermatozoa. (d) End phase with typical spermatozoa detaching. (From W. E. Ankel (1930). Die atypische Spermatogenese von *Janthina* (Prosobranchia, Ptenoglossa). *Z. Zellforsch. mikrosk. Anat.* **11**, 491–608.)

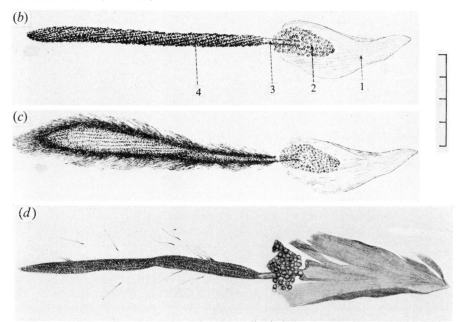

(b)

(c)

(d)

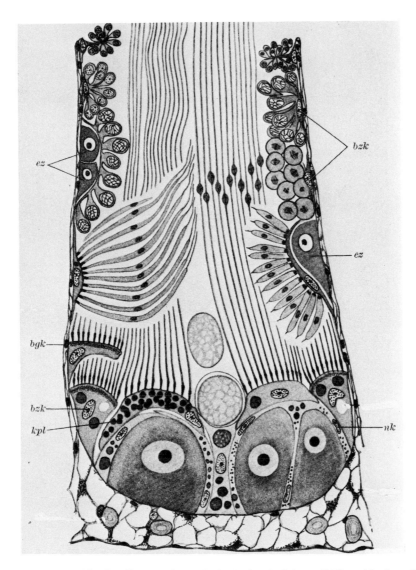

Plate 5. (a) A combination diagram of a testicular 'acinus' of the snail *Planorbis planorbis*. Clones of spermatogonia, spermatocytes and spermatids with their basal cells are seen as they migrate towards the bottom of the acinus. A basal cell near the bottom at left contains droplets of 'Kinoplasma'. At right, the droplets have left the basal cell and have moved along the tails of spermatids. At the bottom are three large oocytes. ez, young oocytes; bzk, basal cell nuclei; bgk, connective tissue nuclei; kpl, droplets of 'Kinoplasma'; nk, nucleus of nurse cell. (From H. Merton (1930). Die Wanderungen der Geschlechtszellen in der Zwitterdrüse von *Planorbis. Z. Zellforsch. mikrosk. Anat.* **10**, 527–51).

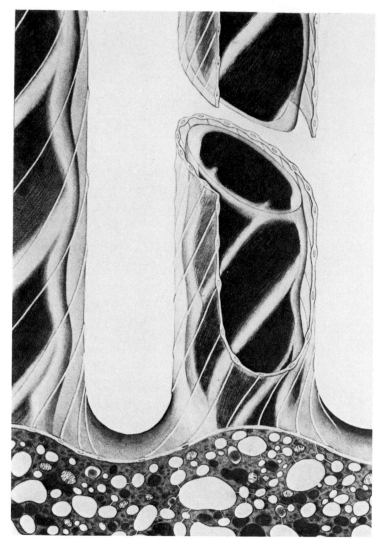

Plate 5. (*b*) Drawing designed to depict the structural relationship of spermatids and nutritive cell in the pond snail, *Cipangopaludina malleata*. The helically coiled head of the spermatids is enveloped by a mantle arising from the nutritive cell. In the mantle a canalicular form of endoplasmic reticulum is seen. The elements run parallel and helical. Vesicular endoplasmic reticulum and mitochondria may also be found in the mantle (see Chapter 5, p. 47). (From G. Yasuzumi, H. Tanaka & O. Tezuka (1960). Spermatogenesis in animals as revealed by electron microscopy. VIII. Relation between the nutritive cells and the developing spermatids in a pond snail *Cipangopaludina malleata* Reeve. *J. biophys. biochem. Cytol.* **7**, 499–504.)

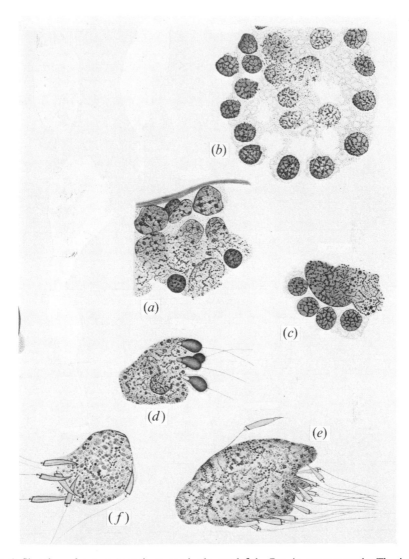

Plate 6. Sketches of spermatogenic stages in the cuttlefish, *Rossia macrosoma* by Thesing (1904). (*a*) Degenerating ('in Auflösung begriffene') spermatogonia; (*b*) degenerating spermatocytes; (*c*) dissolving spermatogonia to which four healthy spermatocytes are applied; (*d*), (*e*) and (*f*) "dissolved" germ cells in various stages. Spermatids are penetrating the "nutritive" mass. This is a remarkable example of an old piece of work in need of reinvestigation. To date it represents a unique finding. (From C. Thesing (1904). Beiträge zur Spermatogenese der Cephalopoden. *Z. wiss. Zool.* **76**, 94–136.)

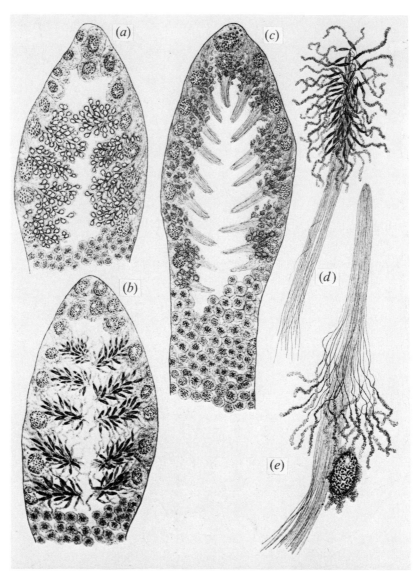

Plate 7. An example of the illustrations of Gilson (1886) showing spermatogenesis in the isopod *Asellus aquaticus*. Successive stages of spermatogenesis are seen in (*a*), (*b*) and (*c*); (*d*) and (*e*) are higher magnifications corresponding to (*b*) and (*c*). For interpretation see in Chapter 6.1, p. 55. (From G. Gilson (1886). Etude comparée de la spermatogenèse chez les arthropodes. *Cellule* **2**, 83–239.)

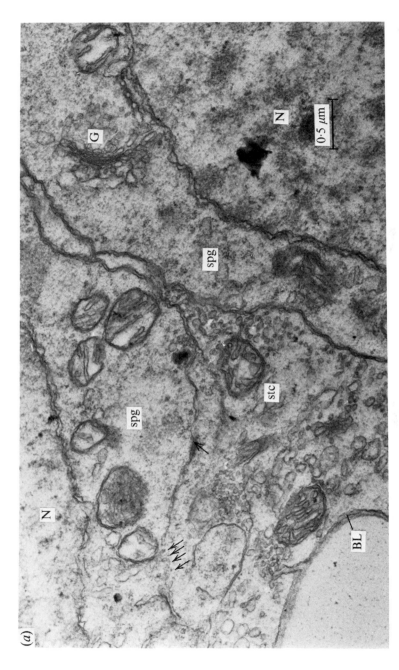

Plate 8. (*a*) Part of the spermatogonial zone of *Diaptomus castor* a calanoid shrimp. A sustentacular cell (stc) surrounds two secondary spermatogonia (spg). N, nuclei; G, developing Golgi. In places the cell membranes of sustentacular and gonial cells are attached by 'petite desmosomes' (single arrow); at other sites the cell membranes are represented only by a row of vesicles (multiple arrows). Fixed in buffered osmium.

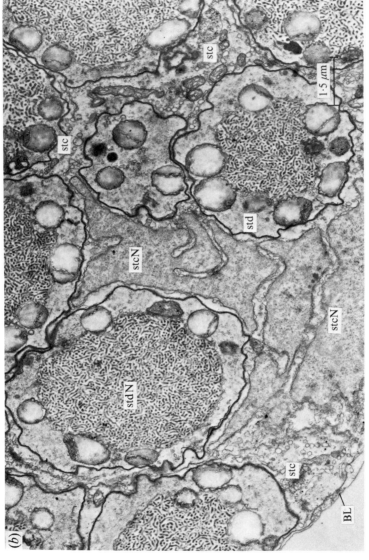

Plate 8. (b) Part of the spermatid zone of *Diaptomus gracialis*. A highly ramified sustentacular cell (stc) with deeply lobed nucleus (stcN) envelops a number of maturing spermatid (std). stdN, nuclei of spermatids. Fixed in buffered osmium. (Both (a) and (b) from B. L. Gupta (1964). Cytological studies of the male germ cells in some freshwater ostracods and copepods. PhD Thesis, Cambridge University.)

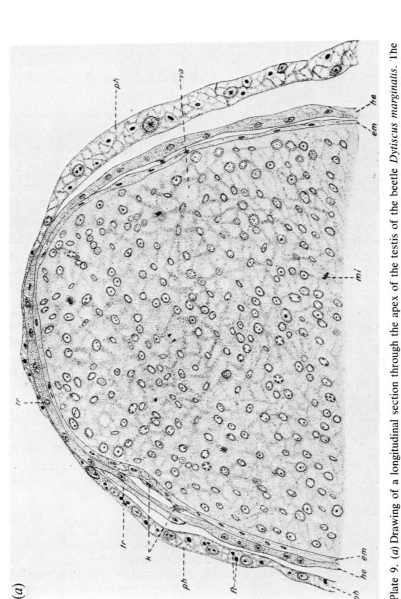

Plate 9. (*a*) Drawing of a longitudinal section through the apex of the testis of the beetle *Dytiscus marginalis*. The diameter of the testis is approx 400 μm. The apex is filled with cells which vary in nuclear size and pattern. Cell borders are indistinct. Cyst formation is not apparent in this region. ph, peritoneal sheath; ft, lipid droplets; tr, trachea; he, testicular lining epithelium; k, nuclei; em, elastic layer; *va*, vacuole; mi, mitosis.

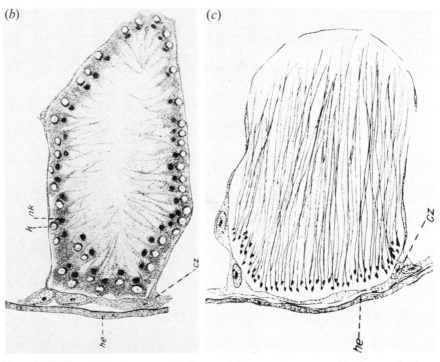

(*b*) (*c*)

Plate 9. (*b*) Cyst of *Dytiscus marginalis* containing maturing spermatids. Long axis about
190 μm. All tails point to the center. k, nucleus; nk, Nebenkern; cz, cyst cells; he, testicular lining epithelium. (*c*) cyst of *Dytiscus marginalis* containing almost mature spermatids
(heads not yet elongate). Long axis approx 220 μm. Cyst wall degenerating in part. All
spermatid heads are applied to intact cyst wall. cz, cyst cells; he, testicular lining epithelium. ((*a*), (*b*) and (*c*) all from C. Demandt (1912). Der Geschlechtsapparat von *Dytiscus
marginalis*. *Z. wiss. Zool.* **103**, 171–299.)

Plate 10. (a) Apical cell from the testis of a caterpillar of *Bombyx mori*. Another representation of the cell depicted by Verson in 1889 (see Fig. 19). me', the place where the apical cell attaches with fine processes to the sheath of the follicle (me); k, nuclei, according to Cholodkovsky nuclei of spermatogonia (sp) partly "'taken in'' by the apical cell; ap, granular cytoplasm of apical cell; nap, the central nucleus of the apical cell. The picture indicates the need for a careful reinvestigation by means of the electron microscope. (From N. Cholodkovsky (1905). Ueber den Bau des Dipterenhodens. *Z. wiss. Zool.* **82**, 389–410.) (b) Sagittal section through the testis of a mature caterpillar of *Tischeria angusticolella. A–D* mark four follicles of the testis. Follicle *C* shows best the successive generations of spermatogonia which develop from the stem cell (U.S.) to disc-shaped clones (Spg.) and spermatocytic cysts (SpC.I). Spt., spermatids; Spt. deg., degenerating spermatids. (From N. Knaben (1931). Spermatogenese bei *Tischeria angusticolella* Dup. *Z. Zellforsch. mikrosk. Anat.* **13**, 291–323.)

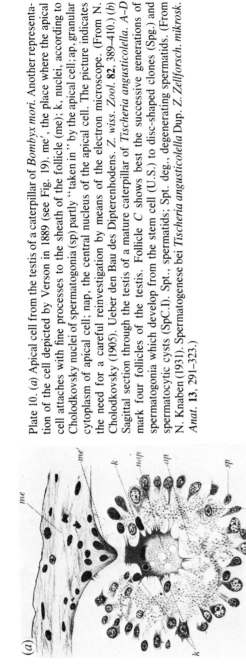

Plate 11. Scanning electron micrograph of spermatogenic clones on the coelomic wall of the polyzoan *Membranopora*. Sg, spermatogonia; Sc, spermatocytes, presumably primary; St, spermatids (see Fig. 21). (Contributed by R. E. Waterman.)

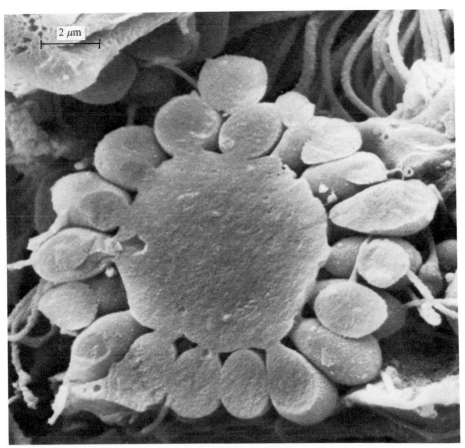

Plate 12. Scanning electron micrograph of a frozen fractured clone of early spermatids of the polyzoan *Membranopora*. The cells can be seen to be continuous with a large cytophore at the center. (Contributed by R. E. Waterman.)

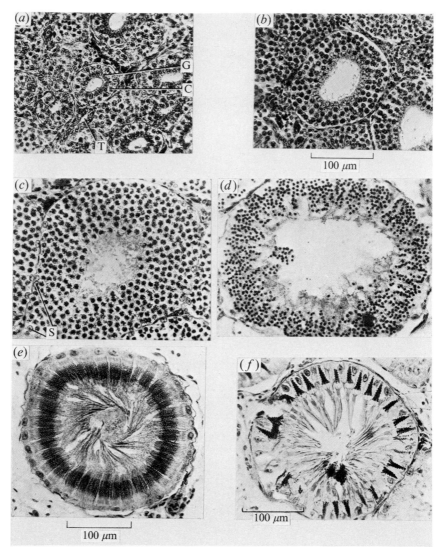

Plate 13. Transverse histological sections of testicular "ampullae" of the basking shark *Cetorhinus maximus* in progressive stages from (*a*) to (*f*). The identifications in the explanations are Matthews'. See Chapter 8, p. 104 for discussion. The layer of cells labeled G in (*a*) and the similar layer in (*b*) are certainly supporting cells ('Sertoli cells' of Matthews) and *not* spermatogonia. G, spermatogonia; C, spermatocytes; S, Sertoli cells; T, a testis tubule in oblique section. In (*c*) the ampulla is largely filled with secondary spermatocytes. (*d*) shows a stage in which spermatids are arranged in groups, presumably in association with individual Sertoli cells. This relationship becomes increasingly clear in (*e*) and (*f*). ((*a*)–(*f*) from L. H. Matthews (1950). Reproduction in the basking shark *Cetorhinus maximus* (Gunner). *Phil. Trans. R. Soc. Lond. Ser. B.* **234**, 247–316.)

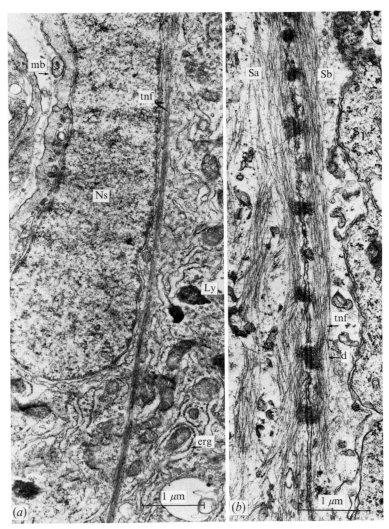

Plate 14. Electron micrographs of the supporting cyst cells ("Sertoli cells") in spermatid cysts of the guppy, *Poecilia reticulata*. (*a*) A bundle of tonofilaments (tnf) runs parallel to the basal lamina (mb) and perpendicular to the plasma membranes, across the longest axis of a "Sertoli cell" surrounding a cyst. An area of cytoplasm along the basal lamina contains hardly any organelles. Ns, nucleus; erg, endoplasmic reticulum. (*b*) Tonofilament bundles (tnf) parallel to the plasma membranes of two bordering "Sertoli cells" (Sa and Sb) around a spermatid cyst (younger than in (*a*)). The fine filaments of desmosomes (d) tie in with the coarser continuous bundles of the filaments running close to the cell surface. ((*a*) and (*b*) from R. Billard (1970). La spermatogenèse de *Poecilia reticulata*. III. Ultrastructure des cellules de Sertoli. *Ann. Biol. anim. Biochim. Biophys.* **10**, 37–50.)

9

Spermatogonia and kinetics of spermatogenesis

Spermatogonia are defined in this book as male germ cells in the first stage of spermatogenesis, the stage of multiplication, or spermatocytogenesis. They develop from cells which variously are called gonocytes, stem cells or protogonia, and which are usually morphologically indistinguishable from the earliest spermatogonia. Studies on the kinetics of mammalian spermatogenesis (see reviews, Chapter 8) have revealed that there are probably always germ cells in the seminiferous tubule which do not take part in the spermatogenic cycle. They lie singly and divide at irregular intervals, apparently independent of the process of spermatogenesis. These cells are stem cells. It is not known under what circumstances they undergo a subtle change in their physiology which causes cell divisions at least in rat and mouse to become incomplete and the daughter cells to remain paired or grouped in syncytial aggregates (Huckins, 1971*a*, *b* and *c*; Oakberg, 1971). At this stage they still divide at irregular intervals. However, at a definite time with respect to the cycle a further change takes place and cell divisions, still incomplete, now occur in strictly regular sequence concomitant with certain stages of the cycle (Fig. 34). With each mitosis the spermatogonia become visibly changed step by step in a line of differentiation which leads to the late spermatogonia, "Type B", which transform into spermatocytes. In its general outline this story is qualitatively and quantitatively well documented in mammals, because it is relatively easy, particularly in whole mounts of tubules (Roosen-Runge, 1955; Clermont & Bustos-Obregón, 1968), to count spermatogonia at each stage of the cycle.

Even in mammals, however, the details of the spermatogonial transformations, particularly the transition from stem cell to spermatogonium are not entirely certain and agreed upon. If it is true, for instance, that this transition is gradual and not abrupt, then it becomes very difficult to define a time at which the first stage of spermatogenesis begins. In non-mammalian animals the data on spermatogonia are scanty and definitions even more uncertain. However, a number of questions may be at least partially answered on the basis of information collected in the review chapters. (1) Do spermatogonia develop in syncytial clones in all animals, and do they always develop synchronously? (2) What is the nature and the number of their sequential cell divisions and is that number significant, particularly with regard to the amount of spermatozoa produced?

Syncytial connections

There is no question that syncytial connections between spermatogenic cells are frequently found throughout the animal kingdom. They have been observed between spermatogonia, spermatocytes and spermatids, and the ultrastructure of the connecting bridges has been repeatedly described, particularly in arthropods and mammals, although there are no detailed studies of the stages of their formation. A brief synopsis of the findings in the main phyla follows. (The numbers which follow the year in reference citations indicate in which chapter's reference list details of the reference are to be found.)

In Porifera the majority of species appear to have syncytial spermatogonia as seen in many light microscopic studies and occasionally in fresh tissues. However, in some species there appear to be no interconnections and in these cases the cells do not develop synchronously. Cases are known in which one of two closely related species shows syncytial, synchronous, the other non-syncytial and asynchronous spermatogonial development (Lévi, 1956; 3) (Fig. 7).

In Coelenterata intercellular bridges between spermatogonia have been repeatedly observed with the electron microscope as well as with the light microscope. According to Brien & Renier-Decoen (1950; 4) the spermatids of *Hydra* always appear separate under the light microscope, but Zihler (1972; 4) in an ultrastructural study found that 10–20% of the spermatocytic divisions are incomplete and result in connected spermatids. It is of interest that Lentz (1966; 9) observed syncytial, synchronous development in the interstitial cells of *Hydra*. These cells are generally believed to give rise to spermatogonia. The situation reminds one somewhat of the behavior of the mammalian gonocytes in their first transformation into paired or aligned spermatogonia described above. On the other hand, Hanisch (1970; 4) reported that in *Eudendrium* male germ cells as they arise from interstitial cells never show intercellular bridges, but spermatocytes are syncytial.

In most Platyhelminthes, as also in Aschelminthes, Polyzoa and Annelida syncytial spermatogonial clones develop in a particular modification, the cytophore (p. 165). However, there are some exceptions. The best described is the parasitic trematode *Schistosomum haematobium* (Lindner, 1914; 7) in which spermatogonia divide completely, remain separate and apparently develop asynchronously. In general, the syncytial clusters of flatworms appear to be remarkably fragile and often break apart at some stage, a behavior which may well occur in other animals in which it has not been traced so clearly.

In Mollusca, Arthropoda and Vertebrata, which together represent almost 90% of animal species, syncytial clones of spermatogonia and of later stages have been well established by light and electron microscopy.

Even in the face of the random nature of sampling over the whole extent of the phyla, it is evident that exceptions to the rule of syncytial, clonal and synchronous development of spermatogonia are excessively rare. The very few authenticated cases occur among sponges, coelenterates and flatworms.

There is little doubt that the syncytial development of germ cell clones is fundamental and important in spermatogenesis. However, the question of what advantage is conferred by this feature has not yet found a convincing answer. Certainly, syncytial clones always develop synchronously but why is synchrony necessary, and is direct cytoplasmic continuity a necessary prerequisite for synchrony? It is certainly true that 'when cells are not in close connexion with one another, biological variability is sufficient to preclude a high degree of synchronization' (Cleland, 1951; 8 (amniotes)), but obviously in spermatogenic tissues, the conditions of a 'close connexion' whatever its nature, are often sufficient to promote a high degree of synchronization way beyond the physical reach of cytoplasmic bridges. In the rat, for instance, a centimeter of length of a tubule may be occupied by cells in one stage of spermatogenesis although no more than a few hundred cells are ever syncytially connected with each other (Dym & Fawcett, 1971; 9). It is more plausible that the main advantage of the syncytial connections does lie in the open communication between cells in certain phases of differentiation, and that synchrony is an inevitable companion phenomenon. Conceivably, the stage in which the intercellular communication may be specifically advantageous is that of the haploid condition, i.e. of the spermatid. The effect of disadvantageous genes in this condition might be balanced by the genes of connected cells. In this way spermatids capable of a normal contribution to a zygote, but handicapped by genes which may influence their normal differentiation, may be enabled to stay within the pool of normal spermatozoa. (See also a discussion remark of Gluecksohn-Waelsch after Hancock, 1972; 9). However, the problem is very complex, because it is known that during the differentiation of spermatogonia and spermatocytes the bridges may be transiently closed (Dym & Fawcett, 1971; 9), thus allowing a cell possible opportunity for its own genotype to exercise a specific effect on its phenotype (discussion remark of Beatty, after Fawcett, 1972; 10). It is quite possible that the syncytial connections are specifically changeable in many animals in which investigators have found intercellular bridges only at certain stages and not at others.

The number of spermatogonial divisions

Table 8 lists most of the cases in which the number of divisions during the spermatogonial multiplication stage have been determined accurately or with reasonable probability. It needs to be explained, however, that the results have been obtained by different methods which vary in reliability. The best method is that of direct quantitation of spermatogonia in a series of stages with or without the assistance of [³H]thymidine labeling. This has been applied almost exclusively to vertebrates and a few insects. Extrapolation of the number of mitoses from the number of late spermatogonia or spermatocytes I connected by cytophores or enclosed in cysts, is somewhat less trustworthy because counting of cells in three-dimensional configurations may present difficulties, and it is usually not

TABLE 8. *Number of spermatogonial generations**

Phylum and species	Number	Derived from†	Author(s); Chapter reference
Dicyemida			
Dicyema aegira	1	Tids	Austin, 1964; 7
Porifera			
Grantia compressa	3		Dendy, 1914; 3
Reniera simulans	5?	Gonia	Tuzet, 1930; 3
Cnidaria			
Phialidium gregarium	4	Quant.	Roosen-Runge, unpublished data
Playthelminthes			
Bucephalus	3	Cytes I	Woodhead, 1931; 7
Dicrocoelium lanceolatum	4	Cytes I	Dingler, 1910; 7
Fasciola hepatica	4	Cytes I	Yosufzai, 1952; 7
Gigantocotyle bathycotyle	4		Willmott, 1950; 7
Gorgoderina attenuata	4	Cytes I	Willey & Koulish, 1950; 7
Haematoloechus medioplexus	4	Cytes I	Burton, 1960; 7
Heronimus chelydrae	4	Tids	Guilford, 1955; 7
Megalodiscus temperatus	4		Van der Woude, 1954; 7
Polychoerus	3	Cytes I	Henley, 1974; 7
Schistosomatium douthitii	4	Tids	Nez, 1954; 7
			Woodhead, 1957; 7
Zygocotyle lunata	4	Cytes I	Willey & Goodman, 1951; 7
Nemertina			
Nectomertes	6?	Tids	Coe & Ball, 1920
Polyzoa			
Alcyonidium gelatinosum	3–4?	Tids	Grellet, 1957; 7
Membranopora villosa	5	Tids	Roosen-Runge, unpublished data
Mollusca			
Ampullaria gigas	6?	Gonia	Sachwatkin, 1920; 5
Cepaea nemoralis	5		Serra & Koshman, 1967a; 5
Helisoma trivolvis	4–5		Abdul-Malek, 1954a; 5
Lymnaea stagnalis	5–6	Tids	Archie, 1943; 5
Opelia crenimarginata	7	Gonia	Bulnheim, 1962a; 5
Ostrea lurida	7–10	Cytes I	Coe, 1932; 5
Philomycus carolinianus	7?	Gonia	Kugler, 1965; 5

* The number of spermatogonial mitoses is always one less than that of the generations. The first generation is supposed to consist of one cell, the fourth of eight, the tenth of 512. Five spermatogonial generations produce 16 spermatocytes I and 64 spermatozoa.

† "Quant." means derived from counts of spermatogonial and/or spermatogonial mitoses. "Gonia" means derived from counts of late secondary gonia connected within a clonal group or situated within a cyst. "Cytes I" and "Tids" means derived from counts of groups of the respective cells.

TABLE 8. (*cont.*)

Phylum and species	Number	Derived from†	Author(s); Chapter reference
Annelida			
Branchiobdella astraci	4	Tids	Bugnion, 1906; 5
Eisenia foetida	5	Gonia	Stang-Voss, 1970; 5
Lumbricus agricolae	2, 3, 4, 5, or 6	Tids	Bugnion, 1906; 5
Lumbricus herculeus	6 or 7	Tids	Chatton & Tuzet, 1943; 5
Lumbricus terrestris	6	Tids	Anderson, Weissman & Ellis, 1967; 5
Arthropoda			
Insecta			
Thysanura			
Lepisma domestica	7		Perrot, 1933; 6.2
Orthoptera			
Gomphocerus	7		Eisentraut, 1926; 6.2
Melanoplus differentialis	8	Tids	White, 1955; 6.2
Oedipoda	8		Eisentraut, 1926; 6.2
Acheta domesticus	10	Quant.	Lyapunova & Zoslimovskaya, 1972; 6.2
Plecoptera	7	Tids	Junker, 1923; 6.2
Ptyraptera			
Goniodes stylifer	5	Tids	Perrot, 1934; 6.2
Hemiptera			
Coccids, several species	6		Hughes-Schrader, 1935, 1946; 6.2 Nur, 1962; 6.2
Coleoptera			
Carabus auratus	9, 10	Tids	Bugnion, 1906; 5
Cetonta aurata	8	Tids	Bugnion, 1906; 5
Dytiscus marginalis	8, 9	Tids	Bugnion, 1906; 5
Prionoplus reticularis	6		Edwards, 1961; 9
Trichoptera			
Several species	8		Phillips, 1970; 6.2
Lepidoptera			
Bombyx mori and			
B. mandarina	7	Cytes I	Kawaguchi, 1928; 6.2
Dieliphela euphorbiae	6–7	Tids	Buder, 1917; 6.2
Hyponomeuta cognatella	7	Tids	Bugnion, 1906; 5
Macrothylacia rubi	6–7	Gonia	Hirschler, 1953; 6.2
Solenobia triquetrella	5	Gonia	Ammann, 1954; 6.2
Tischeria angusticollela	6	Tids	Knaben, 1931; 6.2
Diptera			
Several species	6 or 7		Hess & Meyer, 1968; 6.2
Drosophila melanogaster	5	Tids	Baccetti & Bairati, 1965; 6.2
Dacus oleae	7	Tids	Baccetti & Bairati, 1965; 6.2
Crustacea			
Portunion maenadis	6	Cytes I	Tuzet & Veillet, 1944; 6.1

TABLE 8. (*cont.*)

Phylum and species	Number	Derived from†	Author(s); Chapter reference
Arachnida			
Ornithodorus moubata	5		Breucker & Horstmann, 1972; 6.1
Nemastoma lugubre	5–6	Cytes I	Sokolow, 1929*a*; 6.1
Gamasus, 5 species	1–3	Cytes I	Sokolow, 1934; 6.1
Schizocosa crassipes	3	Cytes I	Hard, 1939; 6.1
Vertebrata			
Pisces			
Torpedo marmorata and other species	5	Tids	Stanley, 1966; 8
Squalus caniculus	13	Gonia	Holstein, 1969; 8
Umbra limi	6–7		Foley, 1926; 8
Phoxinus laevis	5–6		Bullough, 1939; 8
Perca flavescens	5–6	Gonia	Turner, 1919; 8
Brachydanio rerio	5–6	Gonia	Ewing, 1972; 8
Sebastodes paucispinus	7?	Cytes I	Moser, 1967; 8
Aequidens portaligrensis	6?		Polder, 1971; 8
Gambusia holbrokii	10–12	Tids	Geiser, 1924; 9
Poecilia reticulata	14	Quant.	Billard, 1969; 8
Amphibia			
Ambystoma	7		Gurwitsch, 1911; 8
Proteus anguineus	7 & 8	Tids	Stieve, 1920; 8
Triton	7		Gurwitsch, 1911; 8
Mammalia			
Bull	6	Quant.	Hochereau-de Reviers, 1970; 8
Ram	5	Quant.	Ortavant, 1958; 8
Gerbil	5	Gonia	Tobias, 1956; 8
Mouse	6	Quant.	Monesi, 1962; 8
Rat	6	Quant.	Clermont & Bustos-Obregón, 1968; 8
Rabbit	4	Quant.	Swierstra & Foote, 1963; 8
Macacus rhesus	5	Quant.	Clermont & Leblond, 1959; 9
Cercopithecus aethiops	5 or 6	Quant.	Clermont, 1969; 9
Man	4?	Gonia	Rowley, Berlin & Heller, 1971; 8

known how many cells are dying during spermatocytogenesis (see Chapter 10). The method of extrapolating from the number of spermatids joined in cysts or bundles may be misleading under certain circumstances when it cannot be excluded that at some stage clones have combined or subdivided. A principal difficulty inherent in all methods is, of course, the inability in most cases to ascertain beyond doubt whether the cell divisions occur exclusively in spermatogonia or whether they may include divisions of stem cells.

In view of the uncertainties of results obtained in a great variety of animals by different investigators over a period of 70 years, it is all the more remarkable that results are essentially consistent. The data in Table 8 demonstrate a range of spermatogonial generations from 1 to 14, or a range of spermatozoa produced by a single primary spermatogonium from 1 to 36,768 (2^{15}). On first sight this range seems, perhaps, large enough to explain the variations in the amount of spermatozoa produced by different animals, but a brief consideration shows that this conclusion is unwarranted. The highest number of spermatogonial generations appears in the guppy. This is a small fish which ejaculates frequently, but in all probability produces relatively few spermatozoa in a lifetime when compared with, for instance, the perch in which only four to five spermatogonial mitoses occur. Whereas the guppy theoretically produces 36,768 spermatozoa per spermatogonium, the perch only produces 64 or 128. Furthermore, the number of spermatogonial divisions are often identical in invertebrate and vertebrate species which have exceedingly different mechanisms of reproduction and rates of sperm production, as for instance five divisions in mealybugs (coccids) and five in the ram, which may also serve as an example in animals of extremely different sizes. Of the approximately 100 species for which data are available, nine have less than three or more than eight generations of spermatogonia, the remainder are in the range of three to eight, i.e. from 16 to 512 spermatozoa produced by each spermatogonium I. The exceptional cases in which no spermatogonial divisions occur are found in a parasitic dictemid of very aberrant character and in a tick. The last case is interesting because it occurs in two species out of five belonging to the same genus. The other three have either one or two divisions.

High numbers of generations are found in the oyster and a ground beetle, in which they occur probably only in *some* of the clones. Still higher numbers have been found in an orthopteran, a shark and in two species of small livebearing fish. The last have a peculiar mode of copulation in which many ejaculations take place daily over many months. One may speculate that the high number of cells per clone is an adaptation which provides an optimal number of spermatozoa in very small portions, in the form of one or very few cysts or their end products, the spermatozeugmata, which may be conveniently transferred to the female in one ejaculation.

The spermatogonial divisions thus play only a minor role in determining the total amount of spermatozoa released. That determination presumably resides in the number of divisions occurring in primordial germ cells and gonocytes before spermatogenesis begins, i.e. in the amount of undifferentiated germinal tissue available at the time when the primary spermatogonia begin to divide. This will be discussed in the last part of the chapter.

Progressive differentiation of spermatogonial generations

This topic I have reviewed previously with much less information available than is on hand in this book (Roosen-Runge, 1969; 2). The best evidence for differentiation in the course of the spermatogonial multiplication comes once again from vertebrates and insects, but this evidence is reinforced by innumerable observations on many other groups of animals. It is well documented that usually spermatogonia become progressively smaller in the course of their divisions. In some cases, instead of regaining their volume after mitosis, the combined volume of the daughter cells grows only to approximately 1.5 times the volume of the mother cell before the next division occurs (e.g. in snakes (Schreiber, 1949; 9), in the ram (Ortavant, 1958; 8), in man (Roosen-Runge & Barlow, 1953; 9)). The relatively large volume of the primary spermatogonium, is sometimes attained in a conspicuous and rapid spurt of growth of the immediately preceding gonocyte (for instance, in the perch; Turner (1919; 8) (Fig. 25), but I have found not a single report of spermatogonia growing in the process of their evolution.

The most prominent changes in the germ cells during the spermatogonial phase occur in the nucleus. In mammals, these have been minutely described and have been used in the rat to characterize the six generations of spermatogonia (e.g. Clermont & Bustos-Obregón, 1968; 8). Generally, in the primary spermatogonia in fixed preparations, the chromatin of interkinetic nuclei is finely distributed, which caused Regaud (1901; 8 (amniotes)) to describe these cells as 'poussiéreuses', i.e. dusty. With each succeeding generation the chromatin becomes more condensed in fewer and larger flakes (Fig. 2) and the cells may be called 'croûtelleuses', i.e. crusty. Although the minutiae of this process vary in different species, the overall tendency of the heterochromatin to condense is almost always unmistakable in light microscopic preparations. An example of a variation on the theme is found in the monkey (Clermont, 1969; 9) and in man (Rowley *et al.*, 1971; 8). In these species what are assumed to be the earliest spermatogonia (A_1, AD) have finely granulated and deeply staining chromatin which causes the nucleus to appear dark in light microscopic preparations. The subsequent generation of cells (A_2, AP) has a pale nucleus with somewhat coarser and less packed granulation which may be called "dusty", and from these cells develop the typical "crusty" spermatogonia of type B.

Changes similar to those in typical vertebrate spermatogonia have been repeatedly described for insects, for instance, by Mohr (1914; 6.2) in a locust. It is of special interest that in many Orthoptera the X-chromosome has been shown to be negatively heterochromatic in the first spermatogonial generations and positive during the last (Hannah-Alava, 1965; 6.2). In other invertebrates the increase in chromatin condensation is inconspicuous or may not be present at all.

The clearest evidence of progressive changes in the nucleus comes from radioautographic analysis of DNA synthesis in spermatogonia of rat and mouse (Hilscher, Hilscher & Maurer, 1966; 9; Monesi, 1962; 8), which showed that the S-phase becomes longer with each successive spermatogonial division and the G_2-phase shorter, although the duration of the cell cycle remains almost constant (26–30 hours in the mouse and 39–42 hours in the rat). In the only comparable study on insects, Lyapunova & Zoslimovskaya (1973; 6.2) apparently found no conspicuous, and certainly no progressive, changes in S- or G_2-phase through 10 generations of spermatogonia with a steady cell cycle of 24 hours. Tobias (1956; 8) made the suggestion based on his studies in the gerbil that the changes from early to late spermatogonia represent a stepwise development of the chromosomes toward the condition of meiosis, but there are not sufficient data available to enable us to accept this idea as a principle of spermatogonial differentiation. However, the total evidence is compelling that the process of spermatogonial multiplication is coupled with a process of differentiation whatever its nature may be. The degenerations occurring non-randomly in spermatogonial stages (Chapter 10) provide additional evidence for the peculiar nature of spermatogonial mitoses.

Duration of spermatogenesis

Table 9 contains most of the available data for the duration of stages of spermato-genesis. The total duration of the process is not indicated in the table, because the exact beginning of spermatogenesis has hardly ever been identified unequivo-cally by the individual authors. In consequence, the data for the spermatogonial stage also are usually uncertain. On the other hand, an approximation to the true duration of the whole process may be obtained in most cases by simply summing up the spermatogonial, spermatocytic and spermatid stages. The range of total duration in this minute sample of twenty species is from approx. 3 days in a jellyfish (*Phialidium gregarium*) to approx. 75 days in man. It must be remembered, however, that in many animals with seasonal cycles of reproduction there are long periods during the less active part of the cycle in which spermatogenesis proceeds slowly to certain stages, and then stops for a varying length of time until the beginning of the active reproductive season. No case of this kind is included in the table, but data are available in the literature, particularly on fishes and amphibians.

In general, the duration of stages shows remarkable uniformity, with the excep-tion of the case of the jellyfish which is discussed separately below. Spermiogenesis is often the most easily measured stage and the data on it are probably the most reliable. In 15 species its duration is between 4 and 23 days. The shorter times occur in an echinoderm, insects and possibly in sponges, and the longer ones in mammals and in one insect. It is of particular interest that the duration of spermatid differentiation appears to be relatively independent of temperature. At approximately 14 °C it takes 4 days in a sea-urchin, 10 days in a worm and 8 in a fish. In the same fish it takes 7 days at 25 °C, an acceleration which may not be statistically significant. On the other hand, at the much higher body temperatures of mammals the duration is invariably relatively long, although the lower figures for mammals (boar and ram) are in the range of some insects kept presumably at lower temperatures. The meiotic (cytic) stage shows a range from 4 days (*Urechis* and *Drosophila*) to 28 days (man). In mammals it usually has a relatively long duration, but in several insects it is as long as in some mammals. On average the spermatocytic stage is the longest, but similar in length to the spermatid stage. The spermatogonial stage is apparently almost always the shortest.

In mammals the cycle of the seminiferous tissue, i.e. the interval between two

TABLE 9. *Duration of stages of spermatogenesis in days*

Phylum and species	Gonial stage*	Cytic stage†	Tid stage‡	Temp. (°C)	Author(s); Chapter reference
Porifera					
Axinella damicornis and *A. verrucosa*	←—14 days—→			?	Siribelli, 1962; 3
Polymastia mammilaris	←—14 days—→			?	Sarà, 1961; 3
Cnidaria					
Phialidium gregarium	1?	~ 1	~ 1	15	Roosen-Runge, unpublished data
Echiura					
Urechis caupo	24–25?	3–4	10	12–14	Das, 1968; 7
Arthropoda					
Insecta					
Melanoplus differentialis	8–9	9–10	10		Muckenthaler, 1964; 6.2
Acheta domesticus	14	10–12	17–18		Lyapunova & Zoslimovskaya, 1973; 6.2
Bombyx	4–5	10–12	5–6		Sado, 1961; 9
Drosophila melanogaster	?	4	5		Chandley & Bateman, 1962; 9
Echinodermata					
Strongylocentrotus purpuratus	?	6	4	14	Holland & Giese, 1965; 7
Chordata					
Teleosti					
Oryzias latipes	?	5	7	25	Egami & Hyodo-
	?	12	8	15	Taguchi, 1967; 8
Poecilia reticulata	22?	←—14—→		25	Billard, 1968, 1969; 8
Poecilia shenops	?	←—21—→		?	De Felice & Rasch, 1969; 8
Mammalia					
Boar	7.6?	12.1	14.0		Swierstra, 1968; 8
Dog	11.9?	21.4	21.1		Foote, Swierstra & Hunt, 1962; 8
Macaca arctoides		45			Clermont & Antar, 1973; 8
Man	24–26?	27–28	22–23		Heller & Clermont, 1963; 8
Mouse	7	13–14	14–15		Oakberg, 1956; 8
Rabbit	7.9	17.3	15.6		Swierstra & Foote, 1963; 8
Ram	8–9?	16–17	13–14		Ortavant, 1958; 8
Rat	9.0?	18–19	20–21		Clermont, Leblond & Messier, 1959; 9

* The duration of the spermatogonial stage often remains dubious, because its beginning is difficult to determine.

† The spermatocytic stage comprises the life span of spermatocyte I and II. With most authors it begins with leptotene, but parts or all of the preleptotene phase are sometimes included.

‡ The spermatid stage usually is defined as beginning with the end of the telophase of the second meiotic division and ending with spermiation.

appearances of the same stages of spermatogenesis in a given area of the seminiferous tubule, can be measured with great accuracy. Data on its duration in many species are found in several of the reviews mentioned in Chapter 8 (e.g. Dalcq, 1973). The range is narrow, between 9 days in the ram and 16 days in the human. It is well established that the duration of the cycle is 'species and even strain specific' (Clermont & Harvey, 1965; 8), and varies very little in animals of the same genetic composition. It is most probable that this conclusion is valid for all animals and pertains to all stages of spermatogenesis, although a cycle exists only in relatively few species and data are not yet available to document the conclusion on a broad comparative basis.

The case of *Phialidium gregarium*, a jellyfish, needs some elaboration because of its exceptionally rapid spermatogenesis and of the fact that the results have not been published in detail. The data were obtained through observations on testes severed from animals (Roosen-Runge, 1962; 9). Such organs will survive and continue to release sperm twice daily for 3 days. It was found that after this time the reservoir of young spermatogonia was exhausted and spermatogonia apparently were not renewed under the conditions of the experiment. The advance of the generations of germ cells was observed during the 3 days and it was concluded that the meiotic stage and spermiogenesis both lasted approximately 24 hours. This was confirmed tentatively by labeling with [^{3}H]thymidine. It was considered probable that the spermatogonial stage takes 24 hours at most, although the exact duration could not be determined. While these results were obtained on isolated gonads, there is no reason to assume that the generations behave differently in the intact animal in which a new generation of spermatids can be observed to be released approximately every 12 hours. The singularity of the finding is probably only an accidental result of random sampling, and I believe that other species will be found with similarly short durations of spermatogenesis, particularly among hydromedusae.

The productivity of spermatogenesis

Total sperm production over the lifetime of an animal may be of interest from the point of view of the animal's reproductive biology, but comparisons of absolute productivity become meaningful only when set in relation to the number of eggs available, the routes and conditions of fertilization, the capabilities of the spermatozoa, etc. Quantitative data of this kind are usually very incomplete. In general, spermatozoa are produced in numbers greatly in excess of available eggs, and huge losses may occur in the course of the events which lead to successful fertilization. Rotifera are examples of animals, more frequent than is generally assumed, in which the absolute number of spermatozoa and the ratio of spermatozoa to eggs is low. The number of eggs produced is determined in the embryo and is less than 50, often 10 to 20, in a lifetime (Hyman, 1951; 7). In many species the number of spermatozoa probably does not exceed 100 and may be much less, for instance in *Asplanchna* (J. K. Koehler, personal communication). On the other hand, a rabbit or a man may have a maximum daily output of more than 10^8 spermatozoa, chicken or cattle more than 10^9 and a horse or a pig more than 10^{10} (Amann, 1970; 9). The rat may produce 125,000 spermatids in 1 cm length of tubule during a cycle of 12 days (Roosen-Runge, 1955; 9). At 500 cm tubular length in both testes this should yield 10×10^6 spermatozoa per day. It so happens that a small jellyfish releases about the same amount from its four testes (Roosen-Runge, 1962; 4). The most complete, if tentative, estimate of production over the whole reproductive lifetime of an animal is available for *Drosophila melanogaster* (Tihen, 1946; 9). In this animal primordial germ cells (polar cells) are differentiated at the 256-cell stage of cleavage and of these 9–13 are incorporated into each testis. Tihen postulated that these behave as 'cambial cells', i.e. they give rise at every cell division to one definitive spermatogonium. Throughout the whole fertile period of the fly definitive spermatogonia in four successive divisions produce approximately 150,000 spermatozoa. Tihen calculated that the total number of cell divisions preceding the average spermatozoon may be 122, composed of eight cleavage divisions, 107 cambial (primordial cell, gonocyte or predefinitive spermatogonial) divisions, four definitive spermatogonial divisions and two meiotic divisions.

This example points to the crucial issue of spermatogenic productivity, which as we have seen does not lie in the number of spermatogonial divisions, but in the source of supply residing in pre-spermatogonial cellular generations. Quantitative data on these stem cells are almost non-existent although some could probably be calculated in animals in which the total number of primary spermatogonia can be determined. Qualitatively it is possible to identify a number of patterns by which spermatogonia are supplied and renewed in animals, even though most of these patterns need more rigorous definition in detail:

(1) *Renewal from a persistent extragonadal pool of stem cells.* This occurs, e.g. in coelenterates in various forms. In *Hydra* the stem cells are the interstitial cells in between ectoderm and entoderm, in hydromedusae some characteristic primordial cells usually situated in the entodermal gastrodermis (Plate 2*b*). These cells migrate into the ectodermal gonadal tissue whenever needed. In sponges, stem cells may either be omnipresent in the mesenchymal tissue or may even be created from cells previously differentiated in a different way (see Chapter 3).

(2) *Renewal from intragonadal stem cells* which have become incorporated into the gonad early in development. This is well documented for molluscs, arthropods and vertebrates. There are probably modifications in which stem cells are stored at certain locations within the gonad and enter the germ cell compartments at the beginning of each reproductive period (e.g. in fishes). Stem cells within the germinal compartments may become definitive spermatogonia at various successive times in different areas of the compartments, but in amniotes they transform when neighboring germ cells in the same area have reached certain definitive stages of spermatogenesis. It is this behavior which creates a cycle of the spermatogenic tissue providing a continuous supply of spermatozoa at regular intervals of 9 to 15 days in mammals.

10

Degeneration of germ cells. Polymorphism of spermatozoa. Genetic control

Quantitative parameters alone, such as numbers of spermatogonial generations, the duration of spermatogenic stages (Chapter 9), or the fixed proportion of germ cells and associated nurse cells (Chapter 11), which are all species-specific, indicate genetic control of many events in spermatogenesis. However, this control which is obviously effective and statistically successful, often fails in the case of individual cells or clones. In embryogenesis, cell death is a regular, non-random event (Saunders & Fallon, 1966; 10),* and it is, therefore, not surprising that in the developmental process of sperm formation non-random cellular degenerations have been observed frequently. Furthermore, many species have been shown to produce aberrant or atypical spermatozoa in addition to the typical sperm. These deviations from the average pattern of spermatogenesis possibly indicate something of the nature of the normal control processes and are, therefore, of great interest although certain key points of the phenomena are still very puzzling.

I have recently reviewed data concerning germinal cell loss in normal metazoan spermatogenesis (Roosen-Runge, 1973; 10), and Fain-Maurel (1966; 5) has surveyed polymorphism of sperm in some invertebrate phyla. This chapter is not another review, but a discussion of the topic with the aim of presenting deeper insight into the process of spermatogenesis. The questions that will be asked are: (1) How general is the occurrence of deviations from the average cellular development?; (2) What are the modes of regressive events? When and where in the process do they occur in specific cases?; (3) Are the events species-specific?; and (4) What is their cause and their effect?

Table 10 presents a compilation of well authenticated findings of cellular degenerations and demonstrates the wide distribution of the phenomena. The table does not indicate the true frequency of data which have been contributed, usually incidentally, in the course of descriptions of spermatogenesis.

Pertinent information from many species appears to cover a wide range of phenomena which have in common only the appearance of dying cells. These are usually identified in histological preparations by "pycnotic" fragmented nuclei, cytoplasmic vacuolization, swelling, shrinking, ultrastructural membrane changes, etc. They have been described in single cells or in clones in many stages of spermatogenesis. The table indicates clearly that the occurrence of lethal deviations from the average spermatogenic development is indeed very widespread.

---

* The numbers which follow the year in reference citations indicate in which chapter's reference list details of the reference are to be found.

TABLE 10. *Examples of animals in which degenerations have been regularly found in spermatogenesis*

Phylum and class	Stage of degeneration	Author; Chapter reference
Platyhelminthes		
Cestoda	Cytes I	Child, 1907; 7
Turbellaria	Tids	Czernosvitov, 1931; 10
Turbellaria	Meiosis	Lentati, 1970; 7
Turbellaria	Gonia (and cytes?)	Schleip, 1907; 7
Trematoda	Gonia	John, 1953; 7
Trematoda	Gonia	Guilford, 1955; 7
Aschelminthes		
Nematoda	Meiosis	Schleip, 1912; 10
Nematoda	Meiosis	Tretjakoff, 1905; 10
Nematoda	Gonial mitosis	Fauré-Fremiet 1913; 7
Polyzoa	Tids	Grellet, 1957; 7
Mollusca		
Prosobranchia	Tids	Tuzet, 1930; 5
Prosobranchia	Tids, other stages	Ankel, 1924; 5
Prosobranchia	Tids	Aubry, 1954a; 5
Prosobranchia	Tids	Quattrini, 1958; 5
Prosobranchia	Meiosis	Bulnheim, 1962a; 5
Bivalvia	Many stages	Strogonova, 1963; 10
Cephalopodia	Gonia, cytes, tids	Thesing, 1904; 5
Annelida		
Oligochaeta	All stages	Voigt, 1885; 7
Oligochaeta	Tids	Chatton & Tuzet, 1941; 7
Oligochaeta	All stages	Tuzet, 1945; 7
Arthropoda		
Chilopoda	Gonia	Tönniges, 1902; 6.1
Insecta		
Coleoptera	Early cytes I	Henderson, 1907; 6.2
Coleoptera	Cytes	Demandt, 1912; 6.2
Coleoptera	Gonia	Bonhag & Wick, 1953; 6.2
Lepidoptera	All stages	Ammann, 1954; 6.2
Diptera	Early cytes I, tids	Lomen, 1914; 6.2
Diptera	Tids	Bairati, 1967; 6.2
Vertebrata		
Fishes	Gonia	Billard, 1969; 8
Amphibia	Cytes	Flemming, 1887; 8
Amphibia	Many stages	Champy, 1913; 10
Reptilia	Gonial mitosis, meiosis	Dalcq, 1921; 8
Mammalia	Gonia, cytes, tids, meiosis	Roosen-Runge, 1955; 10
Mammalia	Gonia, cytes	Ortavant, 1958; 8
Mammalia	Gonia	Clermont, 1962; 8
Mammalia	Meiosis	Swierstra & Foote, 1963; 8
Mammalia	Gonia, other stages	Hochereau-de Reviers, 1970; 8
Mammalia	Meiosis	Skakkebaek *et al.*, 1973; 10
Mammalia	Tids	Holstein, 1975; 10

There is plentiful evidence that in many cases the cellular degenerations are not random in distribution, but are species-specific in quality and quantity and occur exclusively or preferentially at certain definite stages of development. This does not mean that random degenerative events can be excluded, but apparently in all quantitatively investigated cases their frequency was relatively low. Peaks of degeneration occur in particular: (1) during the spermatogonial stage where quantitative investigations in vertebrates have shown them in specific generations (Billard, 1969; 8 and Clermont, 1962; 8); (2) frequently during meiosis, particularly in early stages, for instance in flatworms (Lentati, 1970; 7), *Ascaris* (Tretjakoff, 1905; 10), molluscs (e.g. Bulnheim, 1962a; 5) and mammals (Roosen-Runge, 1955; 10, in the rat and Skakkebaek *et al.*, 1973; 10, in man); and (3) during certain phases of spermiogenesis, in particular during the early stages of spermatid differentiation. While the cells degenerating as spermatogonia or spermatocytes usually do not long survive (their remains often are phagocytized by nurse cells), the degenerating spermatids may appear in various abnormal forms in the semen and may often fall into the category of "atypical" spermatozoa which constitute polymorphism. The total cell loss during spermatogenesis from early spermatogonium to spermiation varies in quantity as well as in quality with the species. Accurate data are rare. According to Billard the actual yield of spermatozoa in the guppy is 65% of the theoretical products of spermatogonial and spermatocytic cell divisions. In the rat four spermatogonial divisions result on average in only five "intermediate" spermatogonia instead of the 16 postulated (Clermont, 1967; 8). During meiosis the cell loss in Sprague–Dawley rats is approximately 2% (Roosen-Runge, 1955; 10), but during the same stage the rabbit shows a deficit of 24% (Swierstra & Foote, 1963; 8). In man there is a cell loss of approximately 35% between the early spermatocytic and the late spermatid stage (Barr, Moore & Paulsen, 1971; 10). In the bull (Brendston & Desjardin, 1974; 8) losses in early spermatogonia (Type A) are 33%, in late ones (Type B) 27%, in the first meiotic division 6% and in the second one 20%. In early spermiogenesis cell numbers remain stable. A conservative estimate is that one spermatogonium of Type A gives rise in the average to 28 spermatozoa instead of the 64 calculated from the number of spermatogonial divisions (four). These figures do not include the possibility that large numbers of spermatids may manifest marginal abnormalities only very late in spermiogenesis and are not recognized as being abnormal. It appears probable that the species-specific cell loss during spermatogenesis is quite usually at least 50% but may be considerably higher.

The total cell loss during spermatogenesis is in every case the result of a series of specific degenerative events which occur at several identifiable stages. There are certain critical times at which a proportion of gametes may begin to deteriorate. Meiosis, especially in leptotene, zygotene and diakinesis, is most frequently identified as a critical event, and spermatogonial mitoses, perhaps only certain ones in the sequence, are a close second. In a highly pertinent paper on spermatogenesis in the slow-worm (*Anguis fragilis*), Dalcq (1921; 8) discussed a great variety of

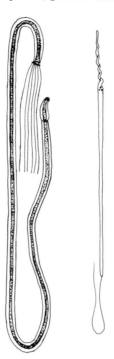

Fig. 35. Typical, eupyrene (left) and atypical apyrene spermatozoon of *Paludina vivipara*, drawn at the same magnification (iron-hematoxylin stained). (After F. Meves (1903). *Arch. mikrosk. Anat.* **61**, 1–84.)

what he called 'atypical gametes' in spermatogonial and spermatocytic stages. He became convinced that all of them originated during mitosis or meiosis, even though some were manifest in interkinetic stages. Dalcq concluded that the most probable cause is an imbalance in the mechanism of division and the distribution of chromosomes, and particularly in the conjugation processes of chromosomes during the meiotic prophases. He assumed that *all abnormal gametes possessed chromosomal abnormalities.* In the unknown processes which produced these, he saw a mechanism rendering lethal certain constitutional faults which, until then, had been compatible with normal development. The underlying mechanisms have not become much clearer since Dalcq's time but they certainly are concerned with gene distribution. According to this concept the non-random degenerations are genetically determined. However, the failures in gene distribution do not always lead to immediate cell death. Often their consequences lead to the phenomenon of polymorphism which is discussed in the following section.

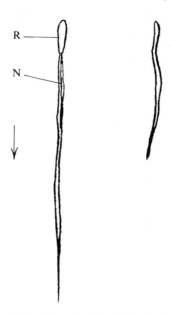

Fig. 36. Typical eupyrene (left, 95 μm long) and atypical, apyrene (right, 45 μm long) spermatozoan of the acoelan flatworm *Otocelis rubropunctata*. N, nucleus, Feulgen positive; R, cytoplasmic residue; → direction of swimming. (After J. Hendelberg (1969). *Zool. Bidr., Uppsala* **38**, 1–50.)

Polymorphism of sperm

The classical case is that of the pond-snail *Paludina* (Fig. 35) in which two types of spermatozoa were discovered by Siebold (1836*b*; 5). Since that time many cases of conspicuous polymorphism have been found in invertebrates and many have been thoroughly investigated as to development and morphology of the atypical spermatozoa. In an acoelan flatworm Hendelberg (1969; 10) found two kinds of spermatozoa, a very peculiar "typical" one with two flagella, and a smaller atypical one which during spermiogenesis casts off its nucleus with residual cytoplasm and remains in the semen as short cytoplasmic rod (Fig. 36). Diminution or lack of chromatin was established as a frequent characteristic of atypical sperm by Meves (1903; 10) who used the term 'eupyrene' for cells with normal chromatin content, 'oligopyrene' for those with subnormal amounts, and 'apyrene' for cells without chromatin. In *Paludina* the atypical spermatozoa are oligopyrene. Chromatin may be absent or occur in abnormal amounts in flatworms, molluscs, annelids, insects and vertebrates; in short in nearly all animals.

In many cases of polymorphism the aberrant types are morphologically very characteristic and sometimes show a behavior which suggests that they have become functional components of the reproductive process although they have never been seen to accomplish normal fertilization. In the rotifer *Asplanchna* atypical spermatids produce immotile 'rod spermatozoa' which are not cells but cellular products of abnormal spermatids, filled with microtubules (Koehler & Birky, 1966; 7). The function of these "rods" is unknown. It may be related to the peculiar intradermal fertilization of these animals. In the spermatophores of freshwater calanoid shrimps two types of spermatozoa are seen, the typical spindle-shaped sperm and large polygonal cells which swell when the spermatophore gets in contact with the water. Their increase in volume exerts a pushing action on the typical sperm during evacuation (Heberer, 1932; 6.1). In one species, *Diaptomus*, Gupta (1964; 6.1) has described the alternating development of two kinds of spermatids, one with and one without a chromatoid body. The first develops into the typical "fertilizing" sperm, the second into "swelling" spermatozoa (see p. 54).

The most remarkable cases of special differentiation occur in molluscs and specifically in snails. Some of these will be described briefly. Most of the data from the comprehensive work of Ankel (1924, 1930, 1933, 1958; 5). In *Viviparus* atypical spermatozoa are vermiform, oligopyrene and multiflagellar. In *Chenopus* they are apyrene and have no free tails although they are motile. In *Vermetus* they are spindle-shaped, contain large yolk platelets, are apyrene, have free tails, but appear completely immotile. In *Strombus* atypical spermatozoa are equipped with undulating membranes and yolk; they are apyrene and motile. In *Conus* they are spherical with two flagella, without yolk, apyrene and non-motile.

Ankel was able to show that in many species the atypical series is recognizable already in spermatogonial stages. In *Bythinia* and closely related species the atypical spermatozoa are morphologically very similar to typical ones, but they are oligo- or hyperpyrene and all of them arise in meiosis. Furrow (1935; 5) described in *Valvata* three different kinds of atypical spermatozoa which departed from normal development at various stages, all postspermatogonial. The first, resulting in an umbrella-shaped head, can be traced back to an atypical anaphase of the second meiotic division, the second, an oligopyrene or apyrene "microsperm" of one-quarter the length of a normal sperm develops early in spermiogenesis, and the third originates later and becomes "macrosperm" with sickle-shaped, hyperpyrene heads. The last occurs with approximately one-tenth the frequency of typical spermatozoa. While the umbrella-type degenerates quickly, the microsperm survives until spermiation, but does not reach the efferent duct. The macrosperm is evacuated together with typical sperm.

By far the most elaborate and, in fact, unique atypical sperm cell is the giant spermatozoon, also called a spermatozeugma, of the marine snail *Janthina* (Ankel, 1930; 5). When fully matured it is approximately 750 μm long, equipped with a large undulating membrane, called a 'driving plate', a mass of yolk granules, and a tail composed of thousands of fused flagella which have arisen through a process of extreme multiplication of centrioles. The surface of the giant tail is modified by longitudinal rows of small depressions into which typical spermatozoa may insert. Thus the atypical spermatozoon serves as a carrier for hundreds of typical ones.

Atypical spermatogonia become identifiable early. In contrast to typical spermatogonia which develop in or near the lumen, the atypical ones remain attached to the wall of the follicle by a cytoplasmic stalk until they are almost as big as a fully grown oocyte. The resulting spermatocytes grow until they have reached approximately 1200 times the volume they possessed as primary spermatogonia and 400 times that of a typical spermatocyte. No meiotic division takes place. The chromatin becomes fragmented and appears to get lost, so that the giant carrier cell seems to be apyrene.

In *Opalia* Bulnheim (1962*a*; 5) found an atypical spermatozoon which is similar to a spermatozeugma but does not seem to function as a carrier. He suspected that the atypical line does not undergo spermatogonial cell divisions while the typical one goes through a series of six mitoses. In consequence, a small number of atypical cells are scattered individually between the synchronous clones of typical cells.

Still another remarkable case is that of *Fusitriton oregonensis*, communicated to me by Daphne E. Williams (unpublished data). This prosobranch has two different, highly differentiated atypical spermatozoa which may each have a specific function. Type A (''carrier'') is a tailless, apyrene cell, possibly a nutrient source for eupyrene sperm. It is about three times as large as the typical sperm, of which it carries up to a hundred on its surface. Approximately one in a hundred spermatogonia gives rise to a Type A cell. Type B ('lancet') is still larger, also apyrene, but does not serve as a carrier. Its ultrastructure indicates vigorous secretory activity. The cell contains two kinds of secretory droplets one of which appears to be mucous. Exocytosis of the droplets was observed. It is suggested that this may modify the highly viscous seminal fluid. Thus, in *Fusitriton* the spermatogonia produce three cell lines which diverge widely in differentiation and may all three contribute to successful completion of the reproductive process.

According to Ankel, the atypical series of spermatozoal development in the hermaphroditic prosobranchs represents a case of very labile determination of differentiation. He assumed that the same primordial cell can give rise typically to either male or female gametes or to nurse cells (see Chapter 5). The primordial cell is potentially capable of developing several different complexes of functional enzyme systems. Its lability causes it to activate either one or the other of its potencies according to conditions at a given time, and under certain conditions male and female characteristics (enzyme systems) may be partially and compatibly activated in various degrees and combinations. Only very few of the resulting cells have become recognizably adapted to new functions, for instance the carrier cell of *Janthina*. In other cases the atypical forms, particularly those containing appreciable amounts of yolk, may contribute to the nourishment of typical spermatozoa. In the majority there is no apparent function and the atypical cells die after a relatively short life. Certainly the prosobranchs offer unique approaches to the cellular and molecular systems which determine the differentiation of the male gamete in all its stages of development.

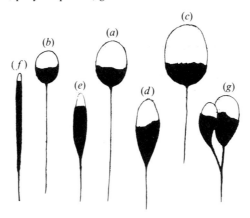

Fig. 37. Major types of spermatozoa in human semen. (*a*) normal; (*b*) small; (*c*) megalo; (*d*) pyriform; (*e*) tapering; (*f*) acutely tapering; (*g*) bicephalic. (After J. MacLeod (1970). *Adv. exp. Med. Biol.* **10**, 481–94.)

It used to be thought that sperm polymorphism occurs only in invertebrates, but this cannot be upheld. As an example, several types of spermatozoa are known to occur in the semen of normal men (MacLeod, 1970; 10) (Fig. 37). The proportion of the major types is reasonably uniform in large groups of men with presumably normal spermatogenesis, for instance macrosperm 3%, microsperm 9%, bicephalic cells 1%. The composition of the population of spermatozoa shows specific individual characteristics and remains quite stable throughout long periods in the life of a man. There is no principal difference between these phenomena and polymorphism of sperm in invertebrates.

It is becoming increasingly clear, however, that polymorphism of spermatozoa may be manifest not only in gross morphological deviations of nuclear and cytoplasmic characteristics, but in simple size differences ('polymegaly', first discussed in annelids by Tuzet, 1930; 5), and in subtle but not random variations in size and configuration of cellular organelles. An example of polymegaly is the difference in sizes of spermatozoa observed in the species group of *Drosophila obscura* (Beatty & Sidhu, 1969; 10). Three types were found (lengths 69 μm, 128 μm and 278 μm) and a possible giant type (430 μm). The relative frequencies of the types are 'virtually constant' in each subspecies, and the pattern of the spermatozoal population was found to be inheritable. There is no information whether all size classes are fertile. Most of the more subtle manifestations of polymorphism, perhaps better called "heterogeneities of sperm", have been investigated by geneticists and have recently been reviewed by Beatty (1970, 1972; 10). Examples are types of mouse spermatozoa characterized by breadth of the head, length of the midpiece, and characteristics of the flagellum or of the acrosome, which all have been shown to be inheritable.

The phenomena of non-random cellular degeneration and of polymorphism and heterogeneity of sperm reveal better than any accounts of normal spermatogenesis the major genetic component in the determination of many facets of the spermatogenic process. They also offer great possibilities for a rigorous quantitative approach to problems of determination. On the other hand, it is well known that chromosomal abnormalities can be produced by a variety of non-genetic influences such as increased temperature, X-rays or chemical interference, and it is often difficult to distinguish between the effects of determining genes and exogenous factors. At the present time, only animals in which the kinetics of spermatogenesis is quantitatively well explored, as in the rat and the fruit fly, lend themselves to critical investigations of the deviations from the normal process.

The loss of germ cells in the course of spermatogenesis is common, regular and often massive, and it raises the question of its *raison d'être*. The phenomenon is apparently closely correlated with the central event of spermatogenesis, meiosis, and with the events which lead up to meiosis, in particular the spermatogonial mitoses. Obviously it is associated in some way with the distribution of genes which regulate phases of the spermatogenic process. Speculations about fundamental reasons for this association are tempting but not pertinent in the present context. However, the general effect of the cell loss is clearly one of winnowing, of "quality control" which serves to select and remove gametes which are in multiple ways not suitable for propagation. By all appearances the effect is large and highly significant, and deserves much more attention than it has received heretofore.

11

Microenvironment of spermatogenesis. Compartments and auxiliary cells

The life task of a male animal is, in a rigorously biological view, the formation of microgametes. Throughout the course of its evolution a species is continually changing in order to adapt to its environment. It may exhibit extreme alterations of its body plan. At the same time the process of sperm formation must remain certain and relatively stable if the race is to survive. Gametogenesis is a major limiting factor in evolution. In view of this, it is of great fundamental interest to explore how much and in what respects the processes of spermatogenesis (and, of course, oogenesis) differ among animals. The data collected in this book have clearly pointed to the fact that the most important clues do not lie in the development of individual germ cells, nor in the mode of spermatogonial divisions, nor in the process of meiosis or the differentiation of the characteristic spermatozoon. They reside in those features by which the developing gamete is related to the body which carries it. The interrelationship of germ cells and soma is the parameter which will be discussed in this chapter, and an attempt will be made to define some simple criteria by which this relationship may be crudely classified in all phyla. Obviously this attempt demands very broad generalizations, and generalizations are fraught with danger.

The great danger...is the temptation to deal with the accepted statements of fact in natural science, as if they might be dealt with deductively, in the same way as propositions in Euclid might be dealt with. In reality, every such statement, however true it may be, is true only relatively to the means of observation and the point of view of those who have enunciated it. So far it may be depended upon. But whether it will bear every speculative conclusion that may be logically derived from it, is quite another question. (Thomas Huxley in a letter concerning Charles Darwin, to Francis Darwin).

In all metazoan bodies, perhaps without exception, spermatogenesis takes place in certain compartments. The germ cells are segregated from the general body tissue by envelopes or capsules which consist essentially of more or less contiguous cells but may also include extracellular elements. There are "primary compartments" which I will define as membrane-bound fluid spaces containing the developing germ cells sometimes together with specialized somatic cells, "secondary compartments" encompassing primary compartments and possibly other cellular elements, and "tertiary compartments" which are defined as encapsulated organs (testes) containing secondary compartments, blood supply and other tissue elements. Primary compartments are of a very limited range of size between 30

TABLE 11. *Terminology used in the literature for spermatogenic sites*
(recommended terms in italics)

Primary compartment*	*Cyst* or *Spermatocyst* (insects, fishes amphibians, etc.) Follicle (sponges, platyhelminthes, teleosts, etc.) Ampulla (selachians) Tubule or *seminiferous tubule* (amniotes) Testicle or testis (sponges, platyhelminthes)
Secondary compartment*	Spermary (coelenterates) Follicle (insects) *Lobule* (fishes) *Tubule* (insects, fishes, amphibians) Ampulla (amphibians) *Testicle* or *testis* (insects, annelids, etc.)
Tertiary (or quaternary) compartment*	*Testicle* or *testis* (all complex gonads)

* For explanation, see text.

and 300 μm in diameter in the whole animal kingdom. Secondary compartments at their smallest may be only a little larger than primary ones, but vary greatly in size. Tertiary compartments or gonads vary by many orders of magnitude. According to Woolley (1975) the seminiferous tubules of five species of marsupials vary in diameter between 0.36 and 0.52 mm. Only one ductule drains these testes and each tubule represents a loop draining into the ductule by two openings. The tubules of marsupials are among the largest primary compartments in the animal kingdom as measured by their diameter.

The spermatogenic compartments have been designated in the literature by a surprising number of different names, the more common of which are listed in Table 11. It will be seen that not only are many terms in use for the same compartments, but the same term is often used for different ones. It is clearly desirable to establish a more uniform and less ambiguous terminology which can be used in comparative considerations and is understood not only by specialists working on certain phyla. The following terms may be acceptable because they consist of old names critically applied and do not include new words. A *cyst* or a *spermatocyst* (and optionally a *follicle*) is a primary compartment of round or ovoid shape. The prototype is the spermatocyst of insects. The seminiferous tubule is a tubular primary compartment, e.g. in amniotes. *Lobule* is a suitable name for a secondary compartment and has been used in this connotation particularly in teleosts. Unfortunately this entity has been commonly called follicle in insects and tubule in many animals in which it happened to have a tubular shape. Secondary compartments which in their entirety represent gonads should, of course, be called *testes*, as for example, in insects with only one testicular lobule. Primary compartments such as the cysts of sponges and flatworms do not deserve the name "testis". Characteristically they contain a single clone of germ cells and are far removed from the complexity of a metazoan "organ". Thus a tapeworm may have thousands of spermatocysts but not thousands of testes.

The immediate microenvironment of the spermatogenic process is the primary compartment, a spermatocyst or a seminiferous tubule in most animals, or the coelomic cavity in Polyzoa, Echiura, Annelida, Pogonophora and probably some other small invertebrate groups. The coelomic type of spermatogenesis is of considerable interest because it appears to be least dependent on the soma. In its usual form germ cells are released into the coelomic fluid from nests of spermatogonia in the coelomic lining, and develop free-floating in clonal groups ("morulae") in the absence of any associated somatic cells. It is characteristic for coelomic clones that they form cytophores, i.e. a common, usually central, cytoplasmic pool to which the cells of a clone adhere with open cytoplasmic connections. The occurrence of cytophores is not entirely confined to the coelomic type of spermatogenesis but wherever they occur there appears to be no close association between germ cells and somatic cells, e.g. in platyhelminthes or nematodes.

In the majority of animals the immediate environment of spermatogenesis is confined within cysts which usually house single clones during their development and in which the fluid environment is presumably modified by an envelope of "cyst cells". In many cases, certainly in sponges and flatworms, these cells show no signs of changes correlated with the differentiation of the germ cells, and there is no evidence of transfer of specific substances. The cyst cells are auxiliary to or supporting spermatogenesis but in such cases there is no reason for calling them "nurse cells". However, in arthropods, in which cystic spermatogenesis is the rule, differentiation of cyst cells is much more complex, and it is correlated more or less closely with the developmental changes in the clone of germ cells which the cyst contains. Excellent examples are the Japanese beetle (p. 34) and the isopod *Asellus* (p. 57). In this state of evolution the cyst cells have quite apparently a nursing function of a kind, even if a direct transfer of nutritive substances has almost never been demonstrated. A good example of a cyst cell which in the course of spermatogenesis develops from a relatively undifferentiated, epithelial cell into a complex supporting cell with functions intimately correlated with the development of the spermatids, is found in anurans (p. 119).

Not all nurse cells are evolved from cyst cells. Examples are the foot- or basal cells of molluscs (Fig. 11) and the "nutritive phagocytes" of echinoderms (p. 99). The concept of a nurse cell is a generalization and encompasses a huge variety of morphological and physiological manifestations which in most cases are only grossly known. An exception is the mammalian Sertoli cell which has revealed a degree of interdependence with germ cells which makes one suspect that phenomena of this kind, if not of the same complexity, must exist in many other cases.

The Sertoli cell, recently the subject of an excellent review (Fawcett, 1975) has not only been much more thoroughly explored than any other nurse or supporting cell, but is characterized by a unique context. It may be defined as a somatic cell associated individually and simultaneously with *several* generations of spermatogenic cells. When this definition is used, a proper Sertoli cell exists almost only in amniotes. In fishes and amphibians the situation is not quite clear. In many teleosts the cyst cells surround only a single clone of germ cells, but in some species a cyst cell may participate in the formation of more than one cyst and thus associate simultaneously with more than one generation of germ cells (R. Billard, personal communication). In amphibians cysts exist during the greatest part of spermatogenesis and the cyst cells are unremarkable, but in many species the cyst cells transform during early stages of spermiogenesis into ramifying nurse cells similar to Sertoli cells in mammals, and the cysts open into the lumen of the tubule so that the *tubule* comes to represent the primary compartment of the spermatids.

There has been a tendency to confer the name of Sertoli cells to cyst cells in amphibians and teleosts and even in crustaceans and molluscs. While in the lower vertebrates the cyst cells have indeed many similarities with Sertoli cells and there are indications that they may be homologues, the ramifying cells of ostracods or the foot cells of pulmonates have only some features in common and the detail of their derivation and relationships to germ cells are so little known that their claim to the name of Sertoli cells is weak, even if, like Sertoli cells, they live in intimate association with germ cells. There is, however, one case of a supporting cell very similar to a Sertoli cell as defined, which occurs in invertebrates. It is the pillar-like, ramifying supporting cell of Hydromedusae (p. 35) which extends through the whole thickness of a seminiferous tissue and is simultaneously and individually in association with several generations of germ cells. The derivation of these cells is completely different from that of Sertoli cells. They are ectodermal, while the Sertoli cells appear to be developed from invaginations of the coelomic epithelium, and the similarities, while startling, have not been explored in detail. At the present state of our knowledge the Sertoli cell in mammals and the supporting cell in jellyfish appear from their morphological interrelationships as the most complex of auxiliary cells – and as a warning of overgeneralizations.

The Sertoli cell in mammals is obviously a highly differentiated unit of surprisingly manifold functions, many of which singly or in various combinations are also seen or at least indicated in other taxa. Thus the Sertoli cell can serve as a model for almost every conceivable property of an auxiliary, somatic cell involved in the spermatogenic process. Such a cell may first of all serve as a skeletal, supporting or sustentacular element in the seminiferous tissue. The elaborate intercellular connections which hold Sertoli cells together (Fig. 34) or the cytoskeleton demonstrable in teleosts (Plate 14) are manifestations of this. Contiguity between cyst or nurse cells is a precondition for their function as barriers between soma and germinal tissue.

The specific locations of intercellular junctions between Sertoli cells (Fig. 34)

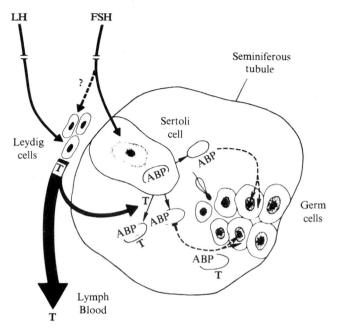

Fig. 38. Diagram of a model of androgen action and the role of the Sertoli cell in the mammalian testis. Testosterone (T) is secreted by Leydig cells stimulated by LH (luteinizing hormone). Most of T is transported out of the testis bound to carrier proteins. Androgen-binding protein (ABP) is produced by Sertoli cells stimulated by FSH (follicle stimulating hormone), and secreted into the seminiferous tubule. Thus an accumulation of androgen is created in close proximity to androgen-dependent germ cells. (After V. Hansson, E. M. Ritzén, F. S. French & S. N. Nayfeh (1975). In *Handbook of Physiology Section 7*, vol. 5 *Male Reproductive System*, ed. D. W. Hamilton & R. O. Greep. Washington, DC: American Physiological Society.)

produce what are in effect two primary compartments within the seminiferous tubules, one containing spermatocytes and spermatids enveloped, except for spermatid tails, by the ramifications of Sertoli cells. It has been shown that certain substances, e.g. lanthanum, penetrate readily into the spermatogonial compartment, but do not pass the Sertoli cell junctions into the central one (for review see Setchell & Waites, 1975). There is no doubt that the two compartments present different microenvironments, although the nature of the differences is still unknown. It appears then that in mammals, as in most animals, the spermatogonial stages are insulated in certain respects from meiotic and spermiogenic stages. However, in all cases in which clones develop separately and singly in primary compartments (e.g. cysts) the insulation of different stages from each other is far more extensive.

The potential role of the Sertoli cell as a mediating element in the control of spermatogenesis is of particular significance. It is now known that Sertoli cells in rats and rabbits produce an androgen-binding protein (ABP) in response to stimulation by follicle-stimulating hormone, and they secrete ABP into the lumen of the seminiferous tubules (Fig. 38) where it presumably combines with testosterone which has passed the soma–germ cell barrier (Hansson, Ritzén, French & Nayfeh, 1975; 11). Even more significant may be the role of the Sertoli cell as an endocrine cell *per se*. It has been suspected for many years that the cell may produce estrogens. Recently this suspicion has been confirmed, at least in tissue culture where it can be shown that follicle-stimulating hormone stimulates synthesis of estradiol-17β (Dorrington & Armstrong, 1975). Conspicuous changes in protein production have been demonstrated in the nurse cells of other phyla (p. 57) but there is no basis as yet for comparative considerations in detail.

There are numerous indications of special functions of auxiliary cells during spermiogenesis. Even in the primitive cysts of sponges in which cyst cells appear as a simple epithelial lining without specific differentiation, the spermatids become oriented and their heads may be applied closely to the cyst cells (Plate 4). In mammals the specific modifications of cytoplasmic organelles near the surface of Sertoli cells where they adjoin the heads of spermatids may be very conspicuous (Plate 14). These morphological features and the characteristic behavior of Sertoli cells in mammals and the supporting cells in amphibians during the release of the spermatids (Fig. 26) seem to reinforce the opinion of Peter (1899, see p. 18) that the main role of the nurse cell is correlated with the peculiar differentiation of the spermatid.

Phagocytosis is a widespread property of supporting cells and wherever it occurs it is usually easily visible. Examples are found throughout the book and range from Platyhelminthes to Mammalia. It is generally understood that the Sertoli cell has strong phagocytic properties and disposes of the residual bodies after spermiation. Fawcett (1975) has pointed out that the evidence for this is somewhat equivocal. The burden of disposal which rests on these cells seems greater than can be accounted for in the electron microscopic demonstration of phagocytosis. This note of caution applies even more to the findings of nuclear and cytoplasmic rests which have been observed so frequently in nurse cells in almost all phyla. On the other hand, it cannot be doubted that nurse cells play a definite role in the process of elimination and, perhaps, recycling of waste products incurred in the process of spermiogenesis as well as in disposal of degenerating spermatogonia and spermatocytes.

The diversities and complexities of the functions of auxiliary cells are not exhausted by the preceding paragraphs. It is reasonable to assume that in the animal kingdom these cells must be not only varied and complex, but also highly species-specific if indeed the spermatogenic process relies on their mediating role for its maintenance in a huge variety of internal environments. Table 12 presents a synopsis of microenvironmental features of spermatogenesis in the animal kingdom as defined by a few gross and oversimplified characteristics. Details and finesse, problems and uncertainties are wiped out in this table except for a few

TABLE 12. *Synopsis of microenvironmental features of spermatogonesis (S.)*
in animals

Dicyemida   Intracellular S., within an axial cell. Unique.

Porifera   Cysts. No nurse cells. No "testes". Synchronous or asynchronous clones. No cytophores. Variable.

Cnidaria   Sites between ectoderm and mesogloeal lamella or in ectoderm. No proper cysts, but auxiliary cells. Nurse cells rare. Cycles of S. Variable.

Platyhelminthes   Cysts, perhaps not always complete. No nurse cells. Cytophores. Clones usually synchronous. Variable.

Nemertina   Cysts. No cytophores. No nurse cells?

Aschelminthes

  Nematoda   Cytophores and rachis (unique). No cysts. No nurse cells. But in *Spirurida* no cytophore, no rachis, but nurse cells.

  Polyzoa   Cytophores attached to coelomic wall. No nurse cells, but epithelial auxiliary cells.

Mollusca   No cysts. No cytophores. Usually complex nurse cells, "foot cells", "basal cells", as carriers of individual clones. Variable.

Echiura   S. in coelom. Cytophores?

Annelida   Later stages of S. in coelom or coelomic pockets. Cytophores. No cysts, except in *Myzostomeria*.

Arthropoda

  Diplopoda   Cysts.

  Chilopoda   Probably no cysts, but nurse cells.

  Insecta   Usually cysts with cyst cells of various degrees of differentiation. In Diptera no true cysts, but germ cell clones more or less segregated by ramified somatic cells. Rarely special nurse cells. Apical cells, unique.

  Crustacea   No true cysts, but germ cells clones more or less segregated by complex somatic cells. Sometimes special nurse cells. Rarely cytophores. Complex and variable.

  Arachnida   Cysts with cyst cells of various degrees of differentiation.

Chaetognatha   S. in coelom. Cytophores?

Pogonophora   S. in coelom. Cytophores.

Echinodermata   "Nutritive phagocytes". No cysts, no cytophores. Scanty data.

Chordata

  Fishes   Cysts with cyst cells often of high complexity. Specific endocrine cells in secondary compartments.

  Amphibians   Cysts. These open and become wall of seminiferous tubule during spermiogenesis. Cyst cells differentiate into complex ramified nurse cells. Endocrine cells in testis.

  Amniotes   Seminiferous tubules. No cysts. Germ cell clones of several generations associated with complex Sertoli cells. Spermatogenic cycle. Endocrine cells in testis.

stark question marks. Yet in its brevity some essential data emerge. (1) It is clear that the process of spermatogenesis almost always involves closely associated somatic cells. (2) The complexity and intimacy of the association in general increases with the evolution of larger bodies and more complex mechanisms controlling homeostasis. (3) Spermatogenesis in the coelomic fluid occurs usually in small animals, although not necessarily in animals of low complexity. While nurse cells are absent in this category, the germ cells are supportive of each other by means of a cytophore. (4) In invertebrates the overwhelming number of species possess spermatogenic cysts. Cyst cells show a wide range of differentiation from simple epithelial cells to large cells of complex morphology. The category of ''spermatogenesis in cysts'' has many subdivisions which are not easily definable at present. (5) It appears that a transition from the cystic to the non-cystic type occurs phylogenetically in the teleosts and amphibians and ontogenetically in amphibians during spermiogenesis. In amniotes, the phenomenon of the spermatocyst does not exist.

The embryological origin of the somatic, auxiliary cells has been investigated in many cases, e.g. in the cyst cells of sponges (p. 25), the nurse of molluscs (p. 39) and the Sertoli cell of mammals. It is clear that there is great diversity. The supporting cells in the jellyfish (p. 35) certainly are transformed epidermal cells. In platyhelminthes the cyst cells are regarded as derived from mesenchymal elements, and in mammals the nurse cells appear to be derived from the coelomic epithelium. When developmental homologies are carefully analyzed the results are often inconclusive because of gaps in the data. The question of whether cyst or nurse cells are derived after the cleavage stage of the egg from the same stem cells as the gametes also has found no unequivocal answer. In sponges and in molluscs there is at least a good possibility that this occurs, and the case is well established for the origin of oocytes and female nurse cells (Raven, 1961, p. 36), but in general the relationship to the germ cell lineage, or to the somatic tissue from which they develop, varies widely.

12

Conclusions

The following conclusions are derived from the data collected and presented in this book.

(1) In Metazoa, male germ cells develop in clones derived from a single, early spermatogonium. The cells of a clone are usually connected in a syncytium through intercellular bridges resulting from incomplete cytoplasmic divisions in mitosis and meiosis. The germ cell syncytium characteristically develops in the spermatogonial stage and usually exists throughout spermatogenesis, but the bridges may not be "open" continuously at all stages. Syncytial clones are synchronous clones. There are rare exceptions of non-syncytial and non-synchronous spermatogenesis in low invertebrate phyla. The full consequences of syncytial development are not well understood at present.

(2) The syncytial clones of male germ cells show a tendency to autonomous development. In the great majority of Metazoa they are isolated from each other in primary compartments, *cysts*, which are surrounded by a barrier-forming membrane of *cyst cells*. There are two major exceptions: (i) The *cytophore*, which connects the cells of a single clone in animals in which spermatogenesis takes its course within the coelom. This structure does not have a delimiting cellular envelope. (ii) The *seminiferous tubules* of higher vertebrates, in which clones are not separated by somatic cells. The seminiferous tubule is a primary compartment in which many syncytial clones live in open communion. In most metazoa multiple primary compartments form secondary and tertiary compartments, i.e. organs, of increasing complexity.

(3) The spermatogenic process characteristically begins with a stage during which spermatogonia undergo differentiation, correlated with a series of special mitoses. There are only two or three cases known in which such a stage of differentiation and multiplication does not precede meiosis.

(4) Meiosis, the spermatocytic stage, is the central event of spermatogenesis. Its nuclear features are essentially the same as in female meiosis, but in the male it results with few exceptions in four equal haploid cells, the spermatids.

(5) The morphological differentiation of the spermatozoon occurs in the spermatid, the postmeiotic, haploid stage. There are rare exceptions in which differentiation begins *during* meiosis. The syncytially connected spermatids, in an "individualization process", separate from each other, lose an appreciable part

of their cytoplasm and become, at least for a short period of time, the most autonomous metazoan cells, spermatozoa. The species-specific features of the spermatozoon determine the events of spermiogenesis but are not appreciably foreshadowed in the spermatogonial and spermatocytic events.

(6) Species specificity of the spermatogenic process is most clearly expressed in features of kinetics, e.g. in the number of spermatogonial mitoses which determines the number of cells in a clone, the duration of developmental stages, and the number of cells which abort in a non-random manner at certain stages of development and are a general characteristic of spermatogenesis. In addition there are, of course, innumerable qualitative indications of species specificity. Genetic control of the process is obviously complex and is exercised at the level of nuclear and cytoplasmic differentiation at many stages. Part of the control may be vested in the mechanism of chromosome distribution.

(7) The process of spermatogenesis is not confined to germ cells but characteristically includes specialized somatic cells. Wherever accessory somatic cells occur they begin their association with the germ cells in the beginning of spermatogenesis. The simplest association is that of the delimiting cyst cell, but the symbiotic relationship between germ cells and soma shows a strong tendency toward increased intimacy and complexity. The Sertoli cell in mammals is the best explored example. This cell has multiple functions, among them its role as a link in the chain of endocrine control. The relatively small proportion of invertebrates in which germ cells are not associated with somatic cells, have cytophores and coelomic spermatogenesis.

(8) Spermatogenesis appears to have been present in its essentials at the very beginning of metazoan evolution. (It may have existed long before that, as it now seems certain that a typical spermatogenic process generating hundreds of thousands of spermatozoa exists in Foraminifera (Bé & Anderson, 1976).) When compared to features such as body size, external body form, or even the diverse phenotypes of spermatozoa, spermatogenesis is a relatively stable and constant phenomenon. It is plausible that the somatic cells which envelop and nurse clones of germ cells, are the mediators between the general soma and the germinal tissue. In consequence, their differentiation and arrangement reflect the phylogenetic position of an animal more clearly than spermatogenesis itself, and may be utilized for a limited and gross, but useful, classification of types of spermatogenesis.

Glossary

In a broadly comparative treatise the lack of a common biological language is a continual trial. Sometimes it is a serious obstacle to understanding. For this reason, an explanation of terms commonly used with different meanings in different contexts is certainly in order, as well as the definition of less well known words. The following Glossary, however, although quite limited in the number of words it contains, has a broader purpose. It is meant to pave the way for a general terminology on which biologists can agree, although it lacks the authority to establish such a terminology with finality. Its aim is to promote a more thorough understanding by explaining the origin and derivation, the applications and limitations of terms in certain contexts. Where I have recommended the use of a term or a definition, it was never because I felt it to be of ultimate correctness, but because it appeared to me the most expedient in the present state of our knowledge. A rigorous linguistic and semantic approach to the language of science may be logically desirable, it may even be the true foundation of science, but it may also detract from the phenomena themselves which form the core of the book.

Some of the items in the Glossary have been discussed repeatedly in the literature, sometimes knowledgeably, but never with definite authority, because an international body for the standardization of biological language does not exist. A number of terms, particularly the stages of spermatogenesis, go back to Vallette St George (1876). His comments are often worth reading. Leblond & Clermont (1952) and Roosen-Runge (1962) have attempted to define terms pertaining to the process of mammalian spermatogenesis. Hannah-Alava (1965) has discussed the terminology of spermatogonia from a somewhat comparative viewpoint. In some cases, for instance those of "nurse cell" or "germinal epithelium", a thorough explanation cannot be attempted in a glossary. Such expressions are important subjects treated throughout the book. When their parameters become clear in a variety of contexts, it also emerges that their definitions in a general biological sense are vague and dubious. In summary, the Glossary is designed to stimulate the conscious, rather than the unconscious, use of terms which, good or bad, are often indispensable.

Abbreviations: Fr., French; Gr., Greek; L., Latin.

Ampulla (L., bottle, i.e. a rounded vessel with a narrow neck) In selachians and amphibians the term is used for the spherical or tubular compartments of the testis, also called 'follicles'. These are connected to the outleading ducts by narrow canals so that the name appears justified on purely morphological grounds. Ampulla should not be confused with "spermatocyst" (see Chapter 11).

Apical cell Used first by Grünberg (1903) for a large cell located at the blind end of the testicular tube or of each testicular "follicle" in *insects* (see Chapter 6.2). The term is also used for the large cell of the covering epithelium found at the proximal end of the testis of nematodes, which is in no way homologous to the apical cell of insects. This large cell, which in male nematodes is supposed to give rise to gametes and in the female to nurse cells, is called "terminal cell".

Apyrene (Gr. *pyren*, nucleus) Without chromatin. Term suggested by Waldeyer to Meves (1903) for certain atypical spermatozoa.

Auxocyte Often used synonymously with *spermatocyte* (also oocyte). Wilson (1925) stated: 'At the end of a certain period division of the gonia ceases and they enter upon a *growth period*, much more prolonged in the female, . . . The cells of this period are accordingly called *auxocytes*'. While this probably holds for all female gametes it does not fit those spermatocytes which show no growth in the period described. For instance, the spermatocytes of many teleosts do not grow at all and should not be designated auxocytes. The spermatocytes of mammals do grow to at least four times the volume of the late spermatogonium, and properly may be called auxocytes.

Basal lamina A layer identifiable with the electron microcope, consisting of amorphous or finely fibrillar material underlying the basal surface of epithelial cells. It is part of the basement membrane.

Basement membrane A microscopic membrane identifiable with the light microscope, underlying epithelia, usually visualized by its collagen content. (See also **Basal lamina**.)

Companion cell The cell which surrounds male germ cells forming a cyst in the testis of teleosts (Ruby & McMillan, 1975). "Cyst cell" is a synonym and preferable because it is self-explanatory.

Cycle of the seminiferous epithelium (see also **Seminiferous epithelium**) In mammals 'a complete series of successive cellular associations appearing in one area of a seminiferous tubule' (Leblond & Clermont, 1952), or more generally 'the series of changes in a given area of seminiferous epithelium between two appearances of the same developmental stages' (Roosen-Runge, 1962). The term was introduced by Regaud (1901) and seems indispensable in the description of kinetics in spermatogenesis. The duration of a cycle is not synonymous with the duration of spermatogenesis. The term is occasionally applied to discontinuous "cycles" with long resting periods such as the annual reproduction cycle of teleosts, but the phenomena in this case are different in detail.

Cyst or spermatocyst (Gr. *kystis*, bladder, sac) A compartment formed by a continuous envelope of cells, "cyst cells", and containing a clone of germ cells. This term is proposed in this book for the most common form of "primary compartment of spermatogenesis" which has in the past been called a host of names in different taxa (see Table 11). (See also **Follicle**.)

Cyst cell A cell which alone or with others forms the wall of a cyst containing male germ cells. The term does not denote any particular function. (See also **Sertoli cell, Lobule boundary cell** and **Nurse cell**.)

Cytophore A central, anuclear mass of cytoplasm by which clones of male germ cells may be united. This arises through incomplete cell division (Stephenson, 1930). It occurs commonly in invertebrates with coelomic spermatogenesis.

Deuterogonium A term used by Hardisty (1965) following the terminology of D'Ancona (1943). Probably synonymous with the first generation of secondary spermatogonia.

Diplonema (Gr. *nēma*, thread and *diploos*, double) The fourth and usually brief stage of prophase of the first meiotic division. The two chromosomes of each homologous pair appear separated and the two chromatids in each chromosome usually are distinguishable. Also known as *diplotene* stage (Gr. *tainia*, ribbon).

Eupyrene (Gr. *pyren*, nucleus) With a normal complement of chromatin. Term suggested by Waldeyer to Meves (1903) for normal spermatozoa.

Follicle (L. *folliculus*, a small bag, a little pod) The term is being used with abandon in histology for small cavities, depressions, "crypts". In the field of reproduction it is most often applied to the "ovarian follicles". In the testis the term is used in lower vertebrates and invertebrates, often in a very confusing way, sometimes for primary, sometimes for secondary spermatogenic compartments. For instance, what is often called a "follicle" in the testis of the shark may be the homologue to a "tubule" in the frog, but what is called a "follicle" in the frog is usually the homologue to a "spermatocyst" in the shark (see Chapter 8 and Table 11). What is called a "follicle" in sponges is in flatworms sometimes called "testicles" or "testes" and should be called "*spermatocysts*" in both. The term is useful and unobjectionable only in its most general meaning. When it is used with specific meaning it should be modified ("ovarian" follicle, "testicular" follicle) and homologies and comparative aspects should be observed.

Germinal epithelium A term which the general biologist should learn to avoid. It was created by Waldeyer (1871) as 'Keimepithel', for the coelomic epithelium covering the gonads of mammals. Waldeyer at that time implied 'somatic origin' of germ cells but retracted this himself in 1906. Germinal epithelium was thought to develop from a thin mesothelial lining into a cuboidal and columnar layer into which germ cells migrate. Felix & Bühler (1906) defined it as 'ein Gemisch von Cölomzellen und Genitalzellen'. But it is doubtful that there is such a mixture in all mammals. In the rat, for instance, (Roosen-Runge, unpublished data) the germ cells are usually found under and not in the germinal epithelium. Consequently the definition has become further diluted to, for instance, 'a modified part of the peritoneal mesothelium' (Hamilton, Boyd & Mossman, 1957). The term has been expanded and used by many investigators for that tissue, generally thought to be epithelial, *in* which or *from* which or *around* which germ cells first differentiate. The use of the term in molluscs (see Chapter 5) is, perhaps the clearest example of the inadequacy of the concept. Although the 'germinal epithelium' in molluscs, whatever it represents, is probably ectodermally derived, although many authors have regarded it as a syncytium and not an epithelium, and although its relationship to germ cells is unclear and apparently varied, the term has been applied again and again. Often it is used carelessly as a synonym for seminiferous epithelium or spermatogenic tissue.

Gonocyte (Gr. *gonos*, begetting or producing) Literally a germinal or stem cell. In a more limited sense it is applied to intragonadal premeiotic cells in mammalian male or female embryos or fetuses. In certain cases, e.g. in the rat (Hilscher *et al.*, 1974) specific terms have been introduced for substages of the "gonocyte stage" such as M-, T_1- and T_2-prospermatogonium.

Leptonema (Gr. *nēma*, thread and *leptos*, thin, fine) The first stage of prophase of the first meiotic division. A network of fine threads of chromosomes in diploid number. Also known as *leptotene* stage (Gr. *tainia*, ribbon).

Limiting membrane A convenient term for the investment of the seminiferous tubule (see entry) and also of the tubules of amphibians and fishes and the follicles of insects. This investment includes at least one, often two layers of basement membrane between which a cellular layer, often myoid in character, is sandwiched.

Lobule boundary cell The term was created by Marshall & Lofts (1956) for 'endocrine' cells in the walls of testicular lobules of teleosts.

Meiosis (Gr. *meiōn*, to diminish) Term first used by Farmer & Moore (1905) (see Chapter 2) for the 'heterotypical' reduction divisions, the central event of gametogenesis.

Metrocyte (Gr. *mētēr*, mother) A rare term for stem cell or spermatogonium. Originally used in French (Gilson, 1886).

Nuage (Fr. cloud) The dense fibrous material (order of magnitude 1 μm or less) seen in germ cells of many animals. It may be the same as "polar granules" in insects or "germinal plasm" in amphibians and has been considered the stuff which determines germ cells. The term was probably first used by André & Rouiller (1957). (For review see Eddy, 1975.)

Nurse cells A general term for somatic cells which appear to support germ cells by some secretory activity or transfer of substances. There is rarely any *direct* evidence for a nursing function, but the term is nevertheless reasonable in many cases because of strong circumstantial evidence. Somatic accessory cells for which there is no evidence of a nursing role should not be called "nursing cells".

Oligopyrene (Gr. *pyren*, nucleus)   With little chromatin. Term suggested by Waldeyer to Meves (1903) for typical spermatozoa with diminished chromatin.

Otheocyte (Gr. *otheo*, thrust or push)   Somatic cell in the spermatogenic acini of decapods (Word & Hobbs, 1958).

Pachenema (Gr. *nēma*, thread and *pakhus*, *thick*)   The third and longest stage of prophase of the first meiotic division. The chromosomes form thick threads in haploid number. This is the stage of *crossing over*. Also known as *pachytene* stage (Gr. *tainia*, ribbon).

Pre-spermatogenesis   A phase in the development of the male germ cells which occurs in the testis and precedes the characteristic spermatogenic phase. The phase is characteristic for mammals and may be undefinable or non-existent in many animals. The term appears to stem from Bouin & Ancel (1926) and was elaborated by Hilscher & Makoski (1968) and Hilscher (1970) for the rat.

Primary compartment   A space containing developing germ cells and often also specialized somatic cells and bounded by a contiguous layer of somatic cells (see Chapter 11). According to this definition the spermatocyst in insects and the seminiferous tubule in mammals are both primary compartments.

Primordial germ cell   A germ cell in its earliest identifiable stage, usually but not necessarily extragonadal. In many cases it is morphologically indistinguishable from a "stem cell".

Prospermatogonium   Forerunner of "differentiating" or of "proliferating" or "definitive" spermatogonia (Table 2) which occurs in mammals in the pre-spermatogenic phase. The term is commonly used in amphibians.

Prospermium   An appropriate name for a spermatozoon (or a "spermatid"?) liberated from the testis in immature condition. The term was used by Oppermann (1935) for the sperm of ticks (see Chapter 6.1). The term lends itself well to the designation of a stage which is functionally immature but often almost indistinguishable morphologically from the mature sperm, as for instance the rabbit spermatozoon before capacitation.

Rachis (Gr. backbone)   In vertebrates this term is used for the vertebral column. In nematodes it is a central strand in the growth zone of the testis to which the spermatogonia are attached. It may be a modified "cytophore".

Seminiferous epithelium   A term used widely for either the aggregate of germ cells or the whole of the tissue in the seminiferous tubules of mammals or the secondary compartments of insects and molluscs. The term is ambiguous. Its noun "epithelium" is not strictly appropriate, and its adjective "seminiferous" may be interpreted as "sperm carrying" (somatic elements) as well as "sperm producing" (germinal elements). I have attempted to avoid the term and have used germinal, seminal or spermatogenic tissue instead.

Sertoli cells   Described first by Sertoli (1865) in the human testis as *celluli ramificati*. The eponym was created by Ebner (1888). Primary definition: the non-germ cells associated with germ cells within the seminiferous tubules of mature mammals. While the prototype exists in the human testis the term may be applied wherever embryological and morphological homologies are firmly established. Synonyms are, in the case of mammals, *supporting*, *sustentacular* or *nurse* cells. For extensive discussion see Chapter 11.

Spermary or **spermiary**   *New Century Dictionary* definition: 'Organ in which spermatozoa are generated'. Turner (1919) used it to denote the testis in teleosts. The term is best reserved for secondary compartments in lower invertebrates, such as *Hydra*.

Spermatoblast   A term which has been used in different contexts with very different meanings, e.g.

   (1) It was used by Ebner (1888, 1902) to designate in the mammalian seminiferous tubule a unit of Sertoli cell and its associated germ cells, in other words a cellular aggregate.

Originally this aggregate was conceived as a syncytium. A revival of the term in this meaning might be useful.

(2) It has been applied to the nurse or supporting cell in the stages of spermiogenesis, predominantly in arthropods (e.g. Gilson, 1886).

(3) It has been used to designate cells giving rise to spermatocytes, i.e. as a synonym for "stem cells" or "spermatogonium" (e.g. Hanisch, 1970, in coelenterates). This would be a logical use for the term, if it were not for the fact that the present terminology for the stages of germ cell development has been established for a century.

Spermatodesm Bundle of zoa held together by a cap of dense periacrosomal material (Szöllösi, 1975). Such bundles are released from cysts into the efferent ducts.

Spermatogonium The term was coined by Vallette St George (1876) for the 'Ursamenzelle'. Its original meaning was that of a "primitive" germ cell. It was not clearly distinguished from a *primordial germ cell*.

A cell in the first stage of spermatogenesis. Spermatogonia are diploid, mitotic and premeiotic. Ultimately spermatogonia take their origin from undifferentiated but sex-determined cells which might be called "protogonia" or "stem cells". (For types of spermatogonia, see Table 2).

Spermatophore Package of spermatozoa surrounded by an external envelope, produced by male accessory organs, and designed to prevent the direct contact of spermatozoa with water. The term is sometimes wrongly used as for spermatogenic cyst (Grier, 1975) or spermatozeugma.

Spermatozeugma (Gr. *zeugma*, union) A term created by Ballowitz (1889), used in two meanings: (1) a vehicle cell for spermatozoa (e.g. molluscs, Ankel, 1930) or (2) bundles of spermatozoa without an external envelope, in insects and fish.

Spermatozoon Term created by Baer (1827). Usually used for the stages of the male gamete after the concentration of nuclear material, but this definition does not permit a clear distinction between a "late" spermatid and an "early" spermatozoon. A better definition would be: The postmeiotic, haploid male germ cell after its release from the seminiferous epithelium (Leblond & Clermont, 1952). In some cases, e.g. in mammals, this definition provides an identifiable beginning of the stage. The definition does not hold in cases where germ cells are developing freely in the coelomic cavity as in certain worms, or where male gametes are transferred to the female in spermatogonial stages as in *Rhizocephala*.

Spermiation The release of spermatozoa from the wall of the seminiferous tubule or, more generally, from the germinal tissue. The term was first used for the toad by Oordt & Klomp (1946).

Spermiogenesis The development of the spermatid, i.e. the postmeiotic, haploid phase of spermatogenesis. In the older German literature, this is called 'Spermiohistogenese' and 'Spermiogenese' is synonymous with the English "spermatogenesis". The synonym "spermioteleosis" (or "spermateleois") is less subject to international confusion.

Spermioteleosis (or **Spermateleosis**) Synonym for spermiogenesis, the final phase of spermatogenesis (Gr. *telos*, the end). The term is more often used for invertebrates, although Bishop & Walton (1960) appear to have established its usage in mammals. For international use the term is preferable to the etymologically unsound "spermiogenesis".

Syngony 'Hermaphroditism is usually called syngony in works on nematodes.' (Hyman, 1951).

Terminal cell In the testis of nematodes. (See **Apical cell.**)

Testicle Literally a *small* testis, but used as a synonym for *testis*.

Testiole An expendable term for a part of a testis, usually synonymous with tubule or, perhaps, lobule (e.g. insects, Hoage & Kessel, 1968).

Testis Derived from the Latin, a seemingly unambiguous term of great antiquity for the male gonad. But in sponges, flatworms and a few others, the term should not be applied

(Chapters 3 and 11). In these animals one should better speak of spermatocysts. In most other animals the testes are *composite organs*, made up of many primary compartments. In view of real controversies, for instance, whether oligochaete 'sperm sacs' are testes or not (Beddard, 1895), a definition is not a luxury but a necessity. An example of misuse of the term is in myriapods (Tönniges, 1902): 'The "testis" consists of three tubes, two of which are to be regarded as "seminal vesicles".'

Tunica albuginea Classical gross anatomical term for the thick connective tissue capsule of the mammalian testis (also used for the ovary), but applicable to any substantial envelope of compact male gonad.

Zygonema (Gr. *nēma*, thread and *zugon*, yoke) Second stage of prophase of the first meiotic division. Homologous chromosomes are becoming applied to each other; a relatively short stage. Also known as *zygotene* stage (Gr. *tainia*, ribbon).

References

Chapter 1. Introduction

Afzelius, B. A. (1972). Sperm morphology and fertilization biology. In *Proceedings of the International Symposium on the Genetics of the Spermatozoon*, ed. R. A. Beatty & S. Gluecksohn-Waelsch, pp. 131–42. Edinburgh: Beatty & Gluecksohn-Waelsch.

Baccetti, B. (1970). *Comparative Spermatology*. New York & London: Academic Press.

Billard, R. (1970). Ultrastructure comparée de spermatozoides de quelques poissons teleostéens. In *Comparative Spermatology*, ed. B. Baccetti, pp. 71–9. New York & London: Academic Press.

Boveri, T. (1899). Die Entwicklung von *Ascaris megalocephala* mit besonderer Berücksichtigung der Kernverhältnisse. *Festschrift 70 Geburtstag v. Kupffer's*. Jena: G. Fischer.

Clermont, Y. & Bustos-Obrégon, E. (1968). Reexamination of spermatogonial renewal in the rat by means of seminiferous tubules mounted "in toto". *Am. J. Anat.* **122**, 237–47.

Dym, M. & Fawcett, D. W. (1970). The blood–testis barrier in the rat and the physiological compartmentation of the seminiferous epithelium. *Biol. Reprod.* **3**, 308–26.

Eddy, E. M. (1974). Fine structural observations on the form and distribution of nuage in germ cells of the rat. *Anat. Rec.* **178**, 731–58.

Eddy, E. M. (1975). Germ plasm and the differentiation of the germ cell line. *Int. Rev. Cytol.* **43**, 229–80.

Fawcett, D. W. (1958). The structure of the mammalian spermatozoon. *Int. Rev. Cytol.* **8**, 195–231.

Fawcett, D. W. (1970). A comparative view of sperm ultrastructure. *Biol. Reprod.* **2**, *Suppl.* 2, 90–127.

Fawcett, D. W. (1975). Ultrastructure and function of the Sertoli Cell. In *Handbook of Physiology, Section 7*, vol. 5 *Male Reproductive System*, ed. D. W. Hamilton & R. O. Greep, chapt. 2, pp. 21–55. Washington, DC: American Physiological Society.

Franzén, A. (1956). On spermiogenesis, morphology of the spermatozoon, and biology of fertilization among invertebrates. *Zool. Bidrag, Uppsala* **31**, 356–479.

Hannah-Alava, A. (1965). The premeiotic stages of spermatogenesis. *Adv. Genet.* **13**, 157–226.

Hilscher, W. (1966). Beiträge zur Orthologie und Pathologie der 'Spermatogoniogenese' der Ratte. *Fortschr. Med.* **84**, 721–3.

Hilscher, W. (1970). Kinetics of prespermatogenesis and spermatogenesis of the Wistar rat under normal and pathological conditions. *Morph. Aspects Androl.* **1**, 17–20.

Hilscher, B., Hilscher, W., Bülthoff-Ohnolz, B., Krämer, U., Birke, A., Pelzer, H. & Gauss, G. (1974). Kinetics of gametogenesis. I. Comparative histological and autoradiographic studies of oocytes and transitional prospermatogonia during oogenesis and prespermatogenesis. *Cell Tiss. Res.* **154**, 443–70.

Hilscher, W. & Makoski, H. (1968). Histologische und autoradiographische Untersuchungen zur 'Präspermatogenese' und 'Spermatogenese' der Ratte. *Z. Zellforsch. mikrosk. Anat.* **86**, 327–50.

Huckins, C. (1971). The spermatogonial stem cell population in adult rats. I. Their morphology, proliferation and maturation. *Anat. Rec.* **169**, 533–58.

Leblond, C. P. & Clermont, Y. (1952). Definition of the stages of the cycle of the seminiferous epithelium of the rat. *Ann. N.Y. Acad. Sci.* **55**, 549–84.

Mahowald, A. P. (1971). Polar granules of *Drosophila*. III. The continuity of polar granules during the life cycle of *Drosophila*. *J. exp. Zool.* **176**, 329–444.

Meves, F. (1907). Die Spermatozytenteilungen bei der Honigbiene (*Apis mellifica* L.), nebst einigen Bemerkungen über Chromatinreduktion. *Arch. mikrosk. Anat.* **70**, 414–91.

Mohr, O. L. (1914). Studien über die Chromatinreifung der männlichen Geschlechtszellen bei *Locusta viridissima*. *Arch. Biol.* (*Liège*) **29**, 579–735.

Monesi, V. (1962). Autoradiographic study of DNA synthesis and the cell cycle in spermatogonia and spermatocytes of mouse testis, using tritiated thymidine. *J. Cell Biol.* **14**, 1–18.

Moses, M. J. (1969). Structure and function of the synaptonemal complex. *Genetics* **61** (*Suppl.*), 41–51.

Moses, M. J., Counce, S. J. & Paulson, D. F. (1975). Synaptonemal complex complement of man in spreads of spermatocytes, with details of the sex chromosome pair. *Science* **187**, 263–5.

Nath, V. (1966). *Animal Gametes* (*Male*). *A Morphological and Cytochemical Account of Spermatogenesis*. London: Asia Publishing House.

Nicander, L. (1967). An electron microscopical study of cell contacts in the seminiferous tubules of some mammals. *Z. Zellforsch. mikrosk. Anat.* **83**, 375–95.

Oakberg, E. F. (1971). Spermatogonial stem cell renewal in the mouse. *Anat. Rec.* **169**, 515–32.

Peter, K. (1899). Die Bedeutung der Nährzelle im Hoden. *Arch. mikrosk. Anat.* **53**, 180–211.

Phillips, D. M. (1974). *Spermatogenesis*. New York: Academic Press.

Regaud, C. (1901). Etudes sur la structure des tubes séminifères et sur la spermatogenèse chez les Mammifères. *Arch. Anat. microsc.* **4**, 101–56, 231–80.

Roosen-Runge, E. C. (1962). The process of spermatogenesis in mammals. *Biol. Rev.*, *Cambridge Phil. Soc.* **37**, 343–77.

Roosen-Runge, E. C. (1969). Comparative aspects of spermatogenesis. *Biol. Reprod.* **1**, 24–39.

Roosen-Runge, E. C. & Leik, J. (1968). Gonocyte degeneration in the postnatal male rat. *Am. J. Anat.* **122**, 275–99.

Sokolow, I. (1934). Untersuchungen über die Spermatogenese bei den Arachniden. V. Über die Spermatogenese der Parasitidae (= Gamasidae, Acarae). *Z. Zellforsch. mikrosk. Anat.* **21**, 42–109.

Tobias, P. V. (1956). *Chromosomes, Sex Cells and Evolution in a Mammal*. London: Percy Lund, Humphries & Co.

Weismann, A. (1892). *Die Entstehung der Sexualzelle bei den Hydromedusen*. Jena: G. Fischer.

Wilson, E. B. (1925). *The Cell in Development and Heredity*. New York: Macmillan.

Chapter 2. Historical Aspects

Baer, C. E. von (1827). Beiträge zur Kenntnis der niederen Thiere. *Nova Acta Leop. Carol.* **13**, 640.

Benda, C. (1887*a*). Zur Spermatogenese und Hodenstruktur der Wirbeltiere. *Anat. Anz.* **2**, 368–70.

Benda, C. (1887*b*). Untersuchungen über den Bau des funktionierenden Samenkanälchens einiger Säugetiere und Folgerungen für die Spermatogenese dieser Wirbeltierklasse. *Arch. mikrosk. Anat.* **30**, 49–110.

Biondi, D. (1885). Die Entwicklung der Spermatozoiden. *Arch. mikrosk. Anat.* **25**, 574–620.

Ebner, V. von (1871). *Untersuchungen über den Bau der Samenkanälchen und die Entwicklung der Spermatozoiden bei den Säugetieren und beim Menschen.* Leipzig.

Ebner, V. von (1888). Zur Spermatogenese bei den Säugetieren. *Arch. mikrosk. Anat.* **31**, 236–92.

Farmer, J. B. & Moore, J. E. S. (1905). On the maiotic phase (reduction divisions) in animals and plants. *Q. J. microsc. Sci.* **48**, 489–558.

Fawcett, D. W. (1975). Gametogenesis in the male and prospects for its control. In *The Developmental Biology of Reproduction.* 33rd Symposium of the Society for Developmental Biology, ed. C. L. Markert & J. Papaconstantinou. New York: Academic Press.

Gilson, G. (1884). Etude comparée de la spermatogénèse chez les Arthropodes: I. *Cellule* **1**, 11–188.

Gilson, G. (1887). Etude comparée de la spermatogénèse chez les Arthropodes: III. *Cellule* **4**, 7–88.

Hyman, L. H. (1940). *The Invertebrates*, vol. 1. New York: McGraw-Hill.

Johnson, A. D., Gomes, W. R. & Vandermark, N. L. (eds.) (1970). *The Testis*, vols. 1, 2 and 3. New York: Academic Press.

Koelliker, R. A. von (1841). Beiträge zur Kenntnis der Geschlechtsverhältnisse und der Samenflüssigkeit wirbelloser Thiere. Berlin.

Koelliker, R. A. von (1856). Physiologische Studien über die Samenflüssigkeit. *Z. wiss. Zool.* **7**, 204–72.

Koelliker, R. A. von (1899). *Erinnerungen aus meinem Leben.* Leipzig: W. Engelmann.

Leblond, C. P. & Clermont, Y. (1952). Definition of the stages of the cycle of the seminiferous epithelium in the rat. *Ann. N.Y. Acad. Sci.* **55**, 548–73.

Lieberkühn, N. (1856). Beiträge zur Entwicklungsgeschichte der Spongillen. *Müller's Arch. Anat. Physiol.* 1–19.

Peter, K. (1899). Die Bedeutung der Nährzelle im Hoden. *Arch. mikrosk. Anat.* **53**, 180–211.

Prenant, A. (1892). Sur la signification de la cellule accessoire du testicule et sur la comparaison morphologique des éléments du testicule et l'ovaire. *J. anat. Physiol.* **28**, 292–321, 529–62.

Retzius, G. (1909). Die Spermien der Crustaceen. *Biol. Untersuch.* **14**, 1–54 (especially p. 24).

Roosen-Runge, E. C. (1952). Kinetics of spermatogenesis in mammals. *Ann. N.Y. Acad. Sci.* **55**, 574–84.

Roosen-Runge, E. C. (1962). The process of spermatogenesis in mammals. *Biol. Rev., Cambridge Phil. Soc.* **37**, 343–77.

Roosen-Runge, E. C. (1969). Comparative aspects of spermatogenesis. *Biol. Reprod.* **1**, 24–39.

Roosen-Runge, E. C. (1974). Die Spermatogenese im Lichte der Evolution. *Verh. anat. Ges.* **68**, 23–37.

Rosenberg, E. & Paulsen, C. A. (eds.) (1970). *The Human Testis.* New York: Plenum Press.

Salensky, W. (1907). Über den Bau der Archianneliden nebst Bemerkungen über den Bau einiger Organe der *Saccacirrus papillocercus* Bobr. *Mem. Acad. Imp. Sci. Petersbourg Ser. V.* **16**, 103–451.

Schulze, F. E. (1875). Bau und Entwicklung von *Sycandra. Z. wiss. Zool.* **25**, *Suppl. 3*, 247–80.

Sertoli, E. (1865). Dell'esistenza di particolari cellule ramificati nei canalicoli seminiferi del testiculo humano. *Morgagni* **7**, 31–9.

Valette St George, la von (1865). Über die Genese der Samenkörper: I. *Arch. mikrosk. Anat.* **1**, 404–14.

Wagner, R. (1836). Die Genesis der Samenthierchen. *Müller's Arch. Anat. Physiol.* 225–31.

Wagner, R. (1839). *Erläuterungstafeln zur Physiologie und Entwicklungsgeschichte.* Leipzig: Voss.

Waldeyer, W. (1887). Bau und Entwicklung der Samenfäden. *Anat. Anz.* **2**, 345–68.

Wilson, E. B. (1925). *The Cell in Development and Heredity.* New York: Macmillan.

Chapter 3. Porifera

Dendy, A. (1914). Observations on the gametogenesis of *Grantia compressa. Q. J. microsc. Sci.* **60**, 313–76.

Eimer, T. (1872). Nesselzellen und Samen bei Seeschwämmen. *Arch. mikrosk. Anat.* **8**, 281–4.

Fell, P. E. (1974). Porifera. In *Reproduction of Marine Invertebrates*, ed. A. C. Giese & J. S. Pearse, vol. 1, pp. 71–5. New York: Academic Press.

Fiedler, K. A. (1888). Über Ei- und Spermabildung bei *Spongilla fluviatalis. Z. wiss. Zool.* **47**, 85–128.

Fincher, J. A. (1940). The origin of germ cells in *Stylotella heliophila* Wilson (Textraxonida). *J. Morph.* **67**, 175–92.

Gatenby, J. B. (1919). Germ cells, fertilization and early development in *Grantia. Linn. Soc. London J. Zool.* **34**, 261–97.

Görich, W. (1904). Zur Kenntnis der Spermatogenese bei den Poriferen und Coelenteraten nebst Bemerkungen über die Oogenese der Ersteren. *Z. wiss. Zool.* **76**, 522–43.

Haeckel, J. (1871). Über die sexuelle Fortpflanzung und das natürliche System der Schwämme. *Jena Z. math. Nat.* **6**, 641–51.

Hyman, L. (1940). Porifera through Ctenophora. In *The Invertebrates*, vol. 3. New York: McGraw-Hill.

Leveaux, M. (1942). Contribution a l'étude histologique de l'ovogénèse et de la spermato-génèse des Spongillidae. *Ann. Soc. zool. Belg.* **73**, 33–50.

Lévi, C. (1956). Etude des Halisarca des Roscoff. Embryologie et systématique des Demo-sponges. *Arch. Zool. exp. gén.* **93**, 1–184.

Lieberkühn, N. (1856a). Beiträge zur Entwicklungsgeschichte der Spongillen. *Müller's Arch. Anat. Physiol.* 1–19.

Lieberkühn, N. (1856b). Addendum to above. *Müller's Arch. Anat. Physiol.* 496–514.

Okada, Y. (1928). On the development of a Hexactinellide sponge, *Farrea sollaisi. J. Fac. Sci. Tokyo Univ.* **2**, 1–27.

Polejaeff, N. (1882). Über das Sperma und Spermatogenesis bei *Sycandra raphanus* Haeckel. *Sitzungsber. Akad. Wiss., Wien* **86**, 276–98.

Sará, M. (1961). Richerche sul gonocorismo ed ermafroditismo nei Porifera. *Boll. Zool.* **28**, 47–60.

Schulze, F. E. (1875). Über den Bau und die Entwicklung von *Sycandra raphanus* Haeckel. *Z. wiss. Zool.* **25**, Suppl. 3, 247–80.

Schulze, F. E. (1877). Untersuchungen über den Bau und die Entwicklung der Spongien. II. Die Gattung *Halisarca. Z. wiss. Zool.* **28**, 1–48.

Schulze, F. E. (1878). Untersuchungen über den Bau und die Entwicklung der Spongien. IV. Die Familie der Aplysinidae. *Z. wiss. Zool.* **30**, 379–420.

Schulze, F. E. (1879). Untersuchungen über den Bau und die Entwicklung der Spongien. VIII. Die Gattungen *Hircinia, Nardo* und *Oligoceras* n.g. *Z. wiss. Zool.* **33**, 1–38.

Schulze, F. E. (1881). Untersuchung über den Bau und die Entwicklung der Spongien. X. *Corticium candelabrum. Z. wiss. Zool.* **35**, 410–30.

Siribelli, L. (1962). Differenze nel ciclo sessuale de *Axinella damicornis* (Esper) ed *Axinella verrucosa* O. Sch. (Demospongiae). *Boll. Zool.* **29**, 319–22.

Tuzet, O. (1930). Sur la spermatogénèse de l'Eponge *Reniera simulans. C. R. Soc. Biol. Paris* **103**, 179–97.

Tuzet, O. (1964). L'origine de la lignée germinale et la gamétogénèse chez les spongaires. In *L'origine de la lignée germinale*, ed. Et. Wolff, pp. 79–111. Paris: Hermann.

Tuzet, O., Garrone, R. & Pavans de Ceccaty, M. (1970). Origine choanocytaire de la lignée germinale male chez le Démosponge *Aplysilla rosea* Schulze (Dendroceratides). *C. R. Acad. Sci., Paris* **270**, 955–7.

Tuzet, O. & Paris, J. (1965). La spermatogénèse, l'ovogénèse, la fecondation et les premiers stades du développement chez *Octavella galangni* Tuzet et Paris. *Vie et Milieu* **15**, 309–27.

Tuzet, O. & Pavans de Ceccaty, M. (1958). La spermatogénèse, l'ovogénèse, la fecondation et les premiers stades du développement d'*Hippospongia communis* LMK. *Bull. Biol. Fr. Belg.* **92**, 331–48.

Valette St George, la von (1865). Über die Genese der Samenkörper. *Arch. mikrosk. Anat.* **1**, 403–14.

Chapter 4. Cnidaria

Aders, W. M. (1903). Beiträge zur Kenntnis der Spermatogenese bei den Cölenteraten. *Z. wiss. Zool.* **74**, 81–108.

Bouillon, J. (1957). Etude monographique du genre *Limnocnida* (Limnomedusae). *Ann. Soc. r. zool. Belg.* **87**, 254–471.

Brien, P. (1966). *Biologie de la reproduction animale.* Paris: Masson & Cie.

Brien, P. & Reniers-Decoen, M. (1950). Etude de *Hydra viridis* (Linnaeus). La blastogénèse, la spermatogénèse, l'ovogénèse. *Ann. Soc. r. zool. Belg.* **81**, 33–110.

Brien, P. & Reniers-Decoen, M. (1951). La gamétogenèse et l'intersexualité chez *Hydra attenuata* (Pallas). *Ann. Soc. r. zool. Belg.* **82**, 285–327.

Campbell, R. D. (1974). Cnidaria. In *Reproduction of Marine Invertebrates*, ed. A. C. Giese & J. S. Pearse, vol. 1, chap. 3. New York: Academic Press.

Chia, F. S. & Rostron, M. A. (1970). Some aspects of the reproductive biology of *Actinia equina* (Cnidaria–Anthozoa). *J. mar. Biol. Assoc., U.K.* **50**, 253–64.

Clark, W. H. & Dewel, W. C. (1974). The structure of the gonads, gametogenesis and sperm–egg interactions in sea anemones (Anthozoa: Actinaria). *Am. Zool.* **14**, 494–510.

Dewel, W. C. & Clark, W. H. (1972). An ultrastructural investigation of spermiogenesis and the mature sperm in the Anthozoan *Bunodosoma cavernata* (Cnidaria). *Ultrastruct. Res.* **40**, 417–31.

Downing, E. R. (1905). The spermatogenesis of *Hydra. Zool. Jahrb.* (*Anat.*) **21**, 379–426.

Franc, J.-M. (1973). Etude ultrastructurale de la spermatogénèse de Ctenaire *Beroe ovata. J. Ultrastruct. Res.* **42**, 255–67.

Hanisch, J. (1972). Die Blastostyle- und Spermienentwicklung von *Eudendrium racemosum* Cavolini. *Zool. Jahrb.* (*Anat.*) **87**, 1–62.

Hertwig, O. & Hertwig, R. (1879). *Die Actinien, anatomisch und histologisch mit besonderer Berücksichtigung des Nervenmuskelsystems.* Jena: Fischer.

Huxley, T. (1849). On the anatomy and the affinities of the family of the Medusae. *Phil. Trans. R. Soc., Lond.* **139**, 425–6.

Kleinenberg, N. (1872). *Hydra. Eine anatomisch-entwicklungsgeschichtliche Untersuchung.* Leipzig: Fischer.

Lyke, E. B. & Robson, E. A. (1975). Spermatogenesis in Anthozoa: differentiation of the spermatid. *Cell Tiss. Res.* **157**, 185–205.

Pianka, H. D. (1974). Ctenophora. In *Reproduction of Marine Invertebrates*, ed. A. C. Giese & J. S. Pearse, vol. 1, chap. 4. New York: Academic Press.

Roosen-Runge, E. C. (1960). Cyclic spermatogenesis in a hydromedusa (*Phialidium*) as compared with the rat. *Anat. Rec.* **136**, 267 (abstract).

Roosen-Runge, E. C. (1962). On the biology of sexual reproduction of hydromedusae, Genus *Phialidium* Leuckhart. *Pac. Sci.* **16**, 15–24.

Roosen-Runge, E. C. (1965). Regeneration of gonads in *Phialidium* (Hydromedusa). In *8th International Congress of Anatomy, Wiesbaden*, p. 101 (abstract). Stuttgart: Georg Thieme.

Roosen-Runge, E. C. (1970). Life cycle of the hydromedusa *Phialidium gregarium* (A. Agassiz, 1862) in the laboratory. *Biol. Bull.* **139**, 203–21.

Roosen-Runge, E. C. & Szollosi, D. (1965). On biology and structure of the testis of *Phialidium* Leuckhart (Leptomedusae). *Z. Zellforsch. mikrosk. Anat.* **68**, 597–610.

Stagni, A. & Lucchi, M. L. (1970). Ultrastructural observations on the spermatogenesis in *Hydra attenuata*. In *Comparative Spermatology*, ed. B. Baccetti, pp. 357–62. New York & London: Academic Press.

Tannreuther, G. W. (1908). The development of *Hydra*. *Biol. Bull.* **16**, 261–74.

Tardent, P. (1974). Gametogenesis in the genus *Hydra*. *Am. Zool.* **14**, 447–56.

Trembley, A. (1744). *Mémoire pour servir à l'histoire d'un genre de polypes d'eau douce à bras en forme de cornes*. Leiden: H. Verbeek.

Zihler, J. (1972). Zur Spermatogenese und Befruchtungsbiologie von *Hydra*. *Roux' Arch.* **169**, 239–67.

Chapter 5. Mollusca

Abdul-Malek, A. T. (1954*a*). Morphological studies on the family Planorbidae (Mollusca: Pulmonata). I. Genital organs of *Helisoma trivolvis* (Say). *Trans. Am. Microsc. Soc.* **73**, 103–24.

Abdul-Malek, A. T. (1954*b*). Morphological studies on the family Planorbidae (Mollusca: Pulmonata). II. The genital organs of *Biomphalaria boissyi*. *Trans. Am. Microsc. Soc.* **73**, 285–96.

Ancel, P. (1903). Histogénèse et structure de la glande hermaphrodite d'*Helix pomatia*. *Arch. Biol.* **19**, 289–652.

Anderson, W. A., Weissman, A. & Ellis, R. A. (1967). Cytodifferentiation during spermiogenesis in *Lumbricus terrestris*. *J. Cell Biol.* **32**, 11–26.

Ankel, W. E. (1924). Der Spermatozoendimorphismus bei *Bythinia tentaculata* L. und *Viviparus viviparus* L. *Z. Zellforsch. mikrosk. Anat.* **1**, 85–166.

Ankel, W. E. (1930). Die atypische Spermatogenese von *Janthina* (Prosobranchia, Ptenoglossa). *Z. Zellforsch. mikrosk. Anat.* **11**, 491–608.

Ankel, W. E. (1933). Untersuchungen über Keimzellenbildung und Befruchtung bei *Bythinia tentaculata* L. II. Gibt es in der Spermatogenese von *Bythinia tentaculata* eine Polymegalie? *Z. Zellforsch. mikrosk. Anat.* **17**, 160–98.

Ankel, W. E. (1958). Beobachtungen und Überlegungen zur Morphogenese des atypischen Spermiums von *Scala clathrus* L. *Zool. Anz.* **160**, 11–12, 261–76.

Ansell, A. D. (1961). The development of the primary gonad in *Venus striatula* (Da Costa). *Proc. malacol. Soc.* **34**, 243–7.

Archie, V. E. (1943). The histology and developmental history of the ovotestis of *Lymnaea stagnalis*. Ph.D. thesis, University of Wisconsin.

Aubry, R. (1954*a*). Les éléments nourriciers dans la glande hermaphrodite de *Limnaea stagnalis* adulte. *C. R. Séanc. Soc. Biol.*, Paris **148**, 1626–9.

Aubry, R. (1954*b*). Le lignée mâle dans la glande hermaphrodite de *Limnaea stagnalis* adulte. *C. R. Séanc. Soc. Biol.*, Paris **148**, 1856–8.

Baba, K. (1937). Contribution to the knowledge of a nudibranch, *Okadeia elegans* Baba. *Jap. J. Zool.* **7**, 147–90.

Battaglia, B. (1954). Contributo alla conoscenza morfologica e citochemica della spermato-genesi nei Prosobranchi. La spermatogenesi tipica in *Murex trunculus* L. *L.R.C. Accad. Lincei* **16**, 527–34.

Brökelmann, J. (1963). Fine structure of germ cells and Sertoli cells during the cycle of the seminiferous epithelium in the rat. *Z. Zellforsch. mikrosk. Anat.* **59**, 820–50.

Bugnion, E. (1906). La signification des faisceaux spermatiques. *Bibl. anat.* **16**, 5–52.

Bulnheim. H. P. (1962a). Untersuchungen zum Spermatozoendimorphismus von *Opalia crenimarginata* (Gastropoda, Prosobranchia). *Z. Zellforsch. mikrosk. Anat.* **56**, 300–43.

Bulnheim, H. P. (1962b). Elektronenmikroskopische Untersuchungen zur Feinstruktur der atypischen und typischen Spermatozoen von *Opalia crenimarginata* (Gastropoda, Proso-branchia). *Z. Zellforsch. mikrosk. Anat.* **56**, 371–86.

Buresch, I. (1912). Untersuchungen über die Zwitterdrüse der Pulmonaten. *Arch. Zellforsch.* **7**, 314–43.

Chatton, E. & Tuzet, O. (1943). Recherches sur la spermatogénèse du *Lumbricus herculeus* Sav. Le nucléole séminal et les modalités de son évolution. *Bull. biol. Fr. Belg.* **77**, 29–61.

Coe, W. R. (1932). Development of the gonads and the sequence of the sexual phases in the California oyster (*Ostrea lurida*). *Bull. Scripps Inst. Oceanogr. Tech. Ser.* **3**, 119–44.

Coe, W. R. (1943). Development of the primary gonad and differentiation of sexuality in *Toredo navalis* and other pelecypod mollusks. *Biol. Bull.* **84**, 178–86.

Crabb, E. D. (1927). The fertilization process in the snail *Lymnaea stagnalis oppressa* Say. *Biol. Bull.* **53**, 67–108.

Creek, G. A. (1951). Reproductive system and embryology of the snail *Pomatius elegans* (Müller). *Proc. Zool. Soc., Lond.* **121**, 599–640.

Creek, G. A. (1953). The morphology of *Acme fusca* (Montagu) with special reference to the genital system. *Proc. malacol. Soc., Lond.* **29**, 228–40.

Duncan, C. J. (1958). The anatomy and physiology of the reproductive system of the freshwater snail *Physa fontinalis* (L.). *Proc. Zool. Soc., Lond.* **131**, 55–84.

Eichardt, H. (1950). Zur Kenntnis der cytologischen Vorgänge an den Geschlechtszellen von *Arianta arbustorum* während ihrer Reifungs- und Reifephase, unter besonderer Berück-sichtigung des jahreszeitlichen Rhythmus. *Z. Zellforsch. mikrosk. Anat.* **35**, 110–46.

Fain-Maurel, M. A. (1966). Acquisitions récentes sur les spermatogénèses atypiques. *Ann. Biol. anim. Biochim. Biophys.* **5**, 513–64.

Furrow, C. L. (1935). Evolution of sex in Mollusca. *Trans. III. State Acad. Sci.* **30**, 5–12.

Gatenby, J. B. (1917). The cytoplasmic inclusions of the germ cells. II. *Helix aspersa. Q. J. microsc. Sci.* **62**, 555–612.

Gatenby, J. B. (1919). The cytoplasmic inclusions of germ cells. IV. Notes on the dimorphic spermatozoa of *Paludina* and the giant germ-nurse cells of *Testacella* and *Helix. Q. J. microsc. Sci.* **63**, 401–43.

Hickman, C. P. (1931). The spermiogenesis of *Succinea ovalis* Say, with special reference to the components of the sperm. *J. Morph.* **51**, 244–89.

Hsiao, S. C. T. (1939). The reproductive system and spermatogenesis of *Limacina* (*Spira-tella*) *retroversa* (Flem.). *Biol. Bull.* **76**, 7–25.

Kemnitz, G. A. Von (1914). Beiträge zu unserer Kenntnis des Spermatozoendimorphismus. *Arch. Zellforsch.* **12**, 567–88.

Kugler, O. E. (1965). A morphological and histochemical study of the reproductive system of the slug *Philomycus carolinianus* (Bosc.). *J. Morph.* **116**, 117–32.

Lanza, E. & Quattrini, D. (1964). Richerche sulla biologia del *Veronicellidae* (*Gastropoda soleolifera*). I. La reproduzione in isolamente individuale di *Vaginulus borellianus* (Colosi) e di *Laevicaulis alte* (Férussac). *Monit. Zool. Ital.* **72**, 93–140.

Lemcke, H. & Wingstrand, K. (1959). The anatomy of *Neopilina galathesa* Lemcke. *Galathea Rep.* **3**, 9–72.

Linke, O. (1933). Morphologie und Physiologie des Genitalapparats der Nordseelittorinen. *Wiss. Meeresunters. Abt. Helgoland* **92**, 1–60.

Loosanoff, V. L. (1937). Development of the primary gonad and sexual phases in *Venus mercenaria* L. *Biol. Bull.* **72**, 389–405.

Luchtel, D. (1972a). Gonadal development and sex determination in pulmonate molluscs. I. *Arion circumscriptus. Z. Zellforsch. mikrosk. Anat.* **130**, 279–301.

Luchtel, D. (1972b). Gonadal development and sex determination in pulmonate molluscs. II. *Arion ater rufus* and *Deroceras reticulatum. Z. Zellforsch. mikrosk. Anat.* **130**, 302–11.

Merton, H. (1924). Lebenduntersuchungen an den Zwitterdrüsen der Lungenschnecken. Ein Beitrag zur Protoplasma- und Spermienbewegung. *Z. Zellforsch. mikrosk. Anat.* **1**, 671–86.

Merton, H. (1926). Die verschiedenartige Herkunft des Kinoplasmas der Samenzellen. Eine Parallele zur Nährstoffversorgung des wachsenden Eis. *Biol. Zentralbl.* **46**, 656–78.

Merton, H. (1930). Die Wanderungen der Geschlechtszellen in der Zwitterdrüse von *Planorbis. Z. Zellforsch. mikrosk. Anat.* **10**, 527–51.

Meves, F. (1903). Über oligopyrene und apyrene Spermien und über ihre Entstehung nach Beobachtungen an *Paludina* und *Apygaera. Arch. mikrosk. Anat.* **61**, 1–84.

Pelluet, D. & Watts, A. H. G. (1951). The cytosome of differentiating cells in the ovotestes of slugs. *Q. J. microsc. Sci.* **92**, 453–61.

Pennypacker, M. I. (1930). The germ cells in the hermaphroditic gland of *Polygyra oppressa. J. Morph.* **49**, 415–53.

Pictet (1891). Recherches sur la spermatogénèse chez quelques invertébrés de la Méditerranée. *Mitt. Zool. Stat. Neapel* **10**.

Platner, G. (1885). Über die Spermatogenese bei den Pulmonaten. *Arch. mikrosk. Anat.* **25**, 564–80.

Quattrini, D. (1958). Due particolarità sulla degenerazione degli ovocite e sulla spermiogenesi nei gasteropodi polmonati. *Boll. Zool.* **25**, 149–53.

Quattrini, D. & Lanza, B. (1964a). Le spherule cinetoplasmatiche ('Kinoplasma-Kugeln' di Merton) delle cellule nutrici in *Vaginulus borellianus* (Colosi) (Moll. Gastropoda, Soleolifera). *Boll. Soc. ital. Biol. sper.* **11**, 911–13.

Quattrini, D. & Lanza, B. (1964b). La consistenza numerica dei gruppi isogeni della linea germinale masohile di *Vaginulus borellianus* (Colosi). *Boll. Soc. ital. Biol. sper.* **11**, 1155–6.

Reinke, E. (1913). Development of the apyrene spermatozoa in *Strombus* and of the nurse-cells in *Littorina. Biol. Bull.* **22**, 319–27.

Sachwatkin, V. A. (1920). Das Urogenitalsystem von *Ampullaria gigas* Spix. *Acta zool. Stockh.* **1**, 67–130.

Schitz, V. (1920). Sur la spermatogénèse chez *Cerithium vulgatum* Brug., *Turitella triplicata* Brocchi (Mediterranea monterosato) et *Bittium reticulatum* Da Costa. *Arch. Zool. exp. gén.* **58**, 489–520.

Serra, J. A. & Koshman, R. W. (1967a). Synchronous divisions in spermatogonial cells of a snail. *Can. J. Genet. Cytol.* **9**, 38–43.

Serra, J. A. & Koshman, R. W. (1967b). Nuclear changes of triptional nature in the differentiation of sperm nurse cells in snails. *Can. J. Genet. Cytol.* **9**, 23–30.

Siebold, C. T. von (1836a). Über die Spermatozoen der Crustaceen, Insecten, Gasteropoden und einiger anderer wirbellosen Thiere. *Müller's Arch. Anat. Physiol.* 13–53.

Siebold, C. T. von (1836b). Fernere Beobachtungen über die Spermatozoen der Thiere. II. Die Spermatozoen der *Paludina vivipara. Müller's Arch. Anat. Physiol.* 232–55.

Sóos, L. (1910). Spermiogenesis von *Helix arbustorum. Ann. Musei. Nation. Hungaris* **8**, 304–43.

Stang-Voss, C. (1970). Zur Entstehung des Golgi-Apparates. Elektronenmikroskopische Untersuchungen an Spermatiden von *Eisenia foetida* (Annelida). *Z. Zellforsch. mikrosk. Anat.* **109**, 287–96.

Thesing, C. (1904). Beiträge zur Spermatogenese der Cephalopoden. *Z. wiss. Zool.* **76**, 94–136.

Tikasingh, E. S. (1962). The microanatomy and histology of the parasitic gastropod, *Comenteroxenos paratichopoli* Tikansingh. *Trans. Am. microsc. Soc.* **81**, 320–7.

Tokeya, O. (1924). Über die Spermiogenese von *Melania libertina* Gld. *Folio Anat. Japan* **2**, 131–66.

Tuzet, O. (1930). Recherches sur la spermatogénèse des Prosobranches. *Arch. Zool. exp. gén.* **70**, 95–229.

Tuzet, O. & Mariaggi, J. (1951). La spermatogénèse de *Physa acuta* Drap. *Bull. Soc. Hist. nat., Toulouse* **86**, 245–51.

Walker, M. & MacGregor, H. C. (1968). Spermatogenesis and the structure of the mature sperm in *Nucella lapillus* (L.). *J. Cell Sci.* **3**, 95–104.

Watts, A. H. G. (1952). Spermatogenesis in the slug, *Arion subfuscus*. *J. Morph.* **91**, 53–77.

Wautier, J., Hernandez, M. L. & Richardot, M. (1966). Anatomie, histologie et cycle vital de *Gundlachia Wautiëri* (Mirolli) (Mollusque Basommatophore). *Ann. Sci. nat. (Zool.) Paris* **8**, 495–566.

Weissensee, H. (1916). Die Geschlechtsverhältnisse und der Geschlechtsapparat bei *Anodonta*. *Z. wiss. Zool.* **115**, 262–335.

Whitney, M. E. (1941). The hermaphroditic gland and germ cells of *Vallonia pulchella* Müll. *Pap. Mich. Acad. Sci. Arts Lett.* **26**, 311–38.

Woods, F. H. (1931). History of the germ cells in *Sphaerium striatinum* (Lam.). *J. Morph.* **51**, 545–95.

Yasuzumi, G. (1962). Spermatogenesis in animals as revealed by electron microscopy. XII. Light and electron microscope studies on spermiogenesis of *Cipangopaludina malleata* Reeve. *J. Ultrastruct. Res.* **7**, 488–503.

Yasuzumi, G., Tanaka, H. & Tezuka, O. (1960). Spermatogenesis in animals as revealed by electron microscopy. VIII. Relation between the nutritive cells and the developing spermatids in a pond snail *Cipangopaludina malleata* Reeve. *J. biophys. biochem. Cytol.* **7**, 499–504.

Chapter 6.1. Crustacea, Arachnida, etc.

Abd-el-Wahab, A. (1957). The male genital system of the scorpion, *Buthus quinquestriatus*. *Q. J. microsc. Sci.* **98**, 111–22.

Bérard, J. J. (1974). Etude ultrastructurale de la spermiogénèse et du spermatozoide chez *Daphnia magna* S. (Entomastracés, Branchiopodes, Cladocères). *Bull. Soc. zool. Fr.* **99**, 723–30.

Binford, R. (1913). The germ cells and the process of fertilization in the crab *Menippe mercenaria*. *J. Morph.* **24**, 147–204.

Boissin, L. (1970). Gametogénèse au cours du développement post-embryonnaire et biologie de la reproduction chez *Hysterochelifer meridianus* (L. Koch) (Arachnides, Pseudoscorpions). Ph.D. thesis, Université de Montpellier, France.

Breucker, H. & Horstmann, E. (1972). Die Spermatogenese der Zecke *Ornithodorus moubata* (Murr). *Z. Zellforsch. mikrosk. Anat.* **123**, 18–46.

Chatton, E. & Tuzet, O. (1941). Sur quelques faits nouveaux de la spermiogénèse de *Lumbricus terrestris*. *C. R. Acad. Sci., Paris* **213**, 373–6.

Cronin, L. E. (1947). Anatomy and histology of the male reproductive system of *Callinectes sapidus* Rathbun. *J. Morph.* **81**, 209–39.

Fasten, N. (1914). Spermatogenesis of the American crayfish *Cambarus virilis* and *C. immunis*. *J. Morph.* **25**, 587–649.

Fasten, N. (1918). Spermatogenesis of the Pacific Coast edible crab *Cancer magister* Dana. *Biol. Bull.* **23**, 277–306.

Fasten, N. (1924). Comparative stages in the spermatogenesis of various *Cancer* crabs. *J. Morph.* **29**, 47–61.

Gilson, G. (1884). Etude comparée de la spermatogénèse chez les arthropodes. *Cellule* **1**, 1–188.

Gilson, G. (1886). Etude comparée de la spermatogénèse chez les arthropodes. *Cellule* **2**, 83–239.

Gilson, G. (1887). Etude comparée de la spermatogénèse chez les arthropodes. *Cellule* **4**, 7–106.

Grobben, C. (1878). Beiträge zur Kenntnis der männlichen Geschlechtsorgane der Dekapoden nebst vergleichenden Bemerkungen über die übrigen Thorakostraten. *Arb. zool. Inst. Univ. Wien* **1**, 57–150.

Gupta, B. L. (1964). Cytological studies of the male germ cells in some freshwater ostracods and copepods. Ph.D. thesis, Cambridge University.

Hard, W. L. (1939). The spermatogenesis of the Lycosid spider, *Schizocosa crassipes* (Walckenaer). *J. Morph.* **65**, 121–53.

Heberer, G. (1932). Untersuchungen über Bau und Funktion der Genitalorgane der Copepoden. I. Der männliche Genitalapparat der calanoiden Copepoden. *Z. mikrosk.-anat. Forsch.* **31**, 250–424.

Horstmann, E. (1970). Der Phragmoplast der Spermatiden von *Spirostreptus spec.* (Myriapoda, Diploda). *Z. Zellforsch. mikrosk. Anat.* **104**, 507–16.

Horstmann, E. & Breucker, H. (1969). Spermatozoon und Spermiohistogenese von *Graphidostreptus spec.* (Myriapoda, Diplopoda). *Z. Zellforsch. mikrosk. Anat.* **96**, 505–20.

Ichikawa, A. & Yagamuchi, R. (1958). Studies on the sexual organization of the Rhizocephala. I. The nature of the 'testes' of *Peltogasterella socialis* Krüger. *Annot. zool. Jap.* **31**, 82–86.

Ichikawa, A. & Yagamuchi, R. (1960). Studies on the sexual organization of the Rhizocephala. II. The reproductive functions of the larval (Cypris) males in *Peltogaster* and *Sacculina*. *Annot. zool. Jap.* **33**, 42–56.

Koelliker, A. (1841). Beiträge zur Kenntnis der Geschlechtsverhältnisse und der Samenflüssigkeit wirbelloser Thiere. Berlin.

Koltzoff, N. K. (1908). Studien über die Gestalt der Zelle. II. Untersuchungen über das Kopfskelett des tierischen Spermiums. *Arch. Zellforsch.* **2**, 1–65.

Komai, T. (1920). Spermatogenesis of *Squilla oratorica* de Haar. *J. Morph.* **23**, 307–34.

Matthews, D. C. (1956). The origin of the spermatophoric mass of the sand crab, *Hippa pacifica*. *Q. J. microsc. Sci.* **97**, 257–68.

Meusy, J. J. (1964). Détermination de la durée de la spermatogénèse d'*Orchestia gamarella* Pallas Crustacé amphipode, par injection de thymidine tritée et autoradiographie. *Arch. Anat. microsc. Morph. exp.* **53**, 253–60.

Meusy, J. J. (1968). Ultrastructure de la zone germinative et des gonies du testicule et de l'ovaire d'*Orchestia gamarella* P. (Crustacé Amphipode). *Ann. Sci. Natur. (Zool.)* ser. 12, **10**, 101–16.

Montalenti, G., Vitagliano, G. & de Nicola. N. (1950). The supply of ribonucleic acid to the male germ cells during meiosis in *Asellus aquaticus. Heredity* **4**, 75–87.

Montefoschi, S. (1952). Richerche sulla funzione delle cellule follicolari e sulla relazione con la spermatogenesi in *Anilocra* (Crust. Isop.). *Caryologia* **4**, 25–43.

Oppermann, E. (1935). Zur Entstehung der Riesenspermien von *Argas columbarum* (Shaw) (reflexus F.). *Z. mikrosk.-anat. Forsch.* **37**, 538–60.

Reger, J. F. (1962). A fine structure study on spermiogenesis in the tick *Amblyomma dissimili* with special reference to the development of motile processes. *J. Ultrastruct. Res.* **7**, 550–65.

180   References: Crustacea, Arachnida, etc.

Reger, J. F. (1963). Spermiogenesis in the tick *Amblyomma dissimili* as revealed by electron microscopy. *J. Ultrastruct. Res.* **8**, 607–21.

Reger, J. F. (1966). A comparative study on the fine structure of developing spermatozoa in the isopod *Oniscus asellus*, and the amphipod, *Orchestoidea* sp. *Z. Zellforsch. mikrosk. Anat.* **75**, 579–90.

Reger, J. F. & Cooper, D. P. (1968). Studies on the fine structure of spermatids from the millipede *Polydesmus* sp. *J. Ultrastruct. Res.* **23**, 60–70.

Reinhard, E. G. (1942). The reproductive role of the complemental males of *Peltogaster*. *J. Morph.* **70**, 239–402.

Retzius, G. (1909). Die Spermien der Crustaceen. *Biol. Untersuch.* **14**, 1–54 (see especially Plate X, figs. 6–8, 13, 14).

Rothschild, Lord (1961). Structure and movement of tick spermatozoa (Arachnida, Acari). *Q. J. microsc. Sci.* **102**, 239–47.

Rothschild, Lord (1965). *A Classification of Living Animals*, 2nd edn. Harlow: Longman.

Schmalz, J. (1912). Zur Kenntnis der Spermatogenese der Ostracoden. *Arch. Zellforsch.* **3**, 407–41.

Sharma, G. P. (1950). Spermatogenesis in the spider, *Plexippus paydulli*. *Res. Bull. East Punjab Univ., Zool.* **5**, 67–80.

Siebold, C. T. von (1836). Ueber die Spermatozoen der Crustaceen, Insecten, Gastropoden und einiger anderer wirbelloser Thiere. *Müller's Arch.* 13–53.

Sokolow, I. (1912). Untersuchungen über die Spermatogenese bei den Arachniden. I. Über die Spermatogenese der Skorpione. *Arch. Zellforsch.* **9**, 399–432.

Sokolow, I. (1926). Untersuchungen über die Spermatogenese bei den Arachniden. II. Über die Spermatogenese der Pseudoskorpione. *Z. Zellforsch. mikrosk. Anat.* **3**, 615–81.

Sokolow, I. (1929*a*). Untersuchungen über die Spermatogenese bei den Arachniden. III. Über die Spermatogenese von *Nemastoma lugubre* (Opiliones). *Z. Zellforsch. mikrosk. Anat.* **8**, 617–54.

Sokolow, I. (1929*b*). Untersuchungen über die Spermatogenese bei den Arachniden. IV. Über die Spermatogenese der *Phalangiiden* (Opiliones). *Z. Zellforsch. mikrosk. Anat.* **10**, 164–94.

Sokolow, I. (1934). Untersuchungen über die Spermatogenese bei den Arachniden. V. Über die Spermatogenese der *Parasitidae* (= Gamasidae, Acarae). *Z. Zellforsch. mikrosk. Anat.* **21**, 42–109.

Tönniges, C. (1902). Beiträge zur Spermatogenese und Oogenese der Myriopoden. *Z. wiss. Zool.* **71**, 328–58.

Tuzet, O. & Manier, J. F. (1953). La spermatogénèse d'*Himantarium gabrielis* Meinert. *Ann. Sci. nat. (Zool.)* **15**, 231–9.

Tuzet, O. & Millot, J. (1937). Recherches sur la Spermiogénèse des *Ixodes*. *Bull. Biol. Fr. Belg.* **71**, 190–205.

Tuzet, O. & Veillet, A. (1944). La spermatogénèse du *Portunion maenadis* Giard. *Arch. Zool. exp. gén.* **84**, 1–9.

Warren, E. (1930). Multiple spermatozoa and the chromosome hypothesis of heredity. *Nature, Lond.* **125**, 973–4.

West, W. R. (1953). An anatomical study of the male reproductive system of a Virginia millipede. *J. Morph.* **93**, 123–75.

Weygoldt, P. (1970). Vergleichende Untersuchungen zur Fortpflanzungsbiologie der Pseudoskorpione: II. *Z. Zool. System* **9**, 241–59.

Word, B. H. Jr & Hobbs, H. H. Jr (1958). Observations on the testis of the crayfish *Cambarus montanus acuminatus* Faxon. *Trans. Am. microsc. Soc.* **77**, 435–50.

Wilson, E. B. (1916). The distribution of the chondriosomes to the spermatozoa in scorpions. *Proc. Nat. Acad. Sci., U.S.A.* **2**, 321–4.

Yamagimachi, R. & Fujimaki, N. (1967). Studies on the sexual organization of Rhizocephala. IV. On the nature of the 'testis' of *Thompsonia*. *Annot. zool. Jap.* **40**, 98–104.

Chapter 6.2. Insecta

Aboim, A. N. (1945). Développement embryonnaire et post-embryonnaire des gonades normales et agamétiques de *Drosophila melanogaster*. *Rev. suisse Zool.* **52**, 53–154.

Ammann, H. (1954). Die postembryonale Entwicklung der weiblichen Geschlechtsorgane in der Raupe von *Solenobia triquetrella* F.R. (*Lepid.*) mit ergänzenden Bemerkungen über die Entwicklung des männlichen Geschlechtsapparates. *Zool. Jahrb. Anat.* **73**, 337–94.

Anderson, J. M. (1950). A cytological and cytochemical study of the testicular cyst cells in the Japanese beetle. *Physiol. Zool.* **23**, 308–16.

Baccetti, B. & Bairati, A. (1965). Indagini sull'ultrastruttura delle cellule germinali maschili in *Dacus oleae* ed in *Drosophila melanogaster* Meig. *Redia* **49**, 1–29.

Bairati, A. (1967). Struttura ed ultrastruttura dess'apparato genitale maschile di *Drosophila melanogaster* Meig. *Z. Zellforsch. mikrosk. Anat.* **76**, 56–99.

Ballowitz, E. (1899). Untersuchungen über die Struktur der Spermatozoen, zugleich ein Beitrag zur Lehre vom feineren Bau der kontraktilen Elemente. Die Spermatozoen der Insekten. (I. Coleopteren). *Z. wiss. Zool.* **50**, 317–407.

Baumgartner, W. J. (1929). Die Spermatogenese bei einer Grille *Nemobius fasciatus*. *Z. Zellforsch. mikrosk. Anat.* **9**, 603–39.

Berlese, A. (1910). Monografia de Myrientomata. *Redia* **6**, 1–182.

Bonhag, P. F. & Wick, J. R. (1953). The functional anatomy of the male and female reproductive systems of the milkweed bug, *Oncopeltus fasciatus* (Dallas) (*Heteroptera*: Lygaeidae). *J. Morph.* **93**, 177–283.

Bowen, R. H. (1922a). Studies on insect spermatogenesis. II. The components of the spermatid and their role in the formation of the sperm in *Hemiptera*. *J. Morph.* **37**, 79–193.

Bowen, R. H. (1922b). Studies on insect spermatogenesis. IV. The phenomenon of polymegaly in the sperm cells of the family *Pentatomidae*. *Proc. Am. Acad. Arts Sci.* **57**, 391–415.

Buder, J. E. (1917). Die Spermatogenese von *Deiliphela euphorbiae* L. *Arch. Zellforsch.* **14**, 26–78.

Cantacuzéne, A.-M. (1968). Recherches morphologiques et physiologiques sur les glandes annexes mâles des orthoptères. III. Modes d'association des spermatozoides d'*Orthoptères*. *Z. Zellforsch. mikrosk. Anat.* **90**, 113–25.

Carson, H. L. (1945). A comparative study of the apical cell of the insect testis. *J. Morph.* **77**, 141–61.

Charlton, H. H. (1921). The spermatogenesis of *Lepisma domestica*. *J. Morph.* **35**, 381–423.

Cholodkovsky, N. (1905). Ueber den Bau des Dipterenhodens. *Z. wiss. Zool.* **82**, 389–410.

Cooper, K. W. (1950). Normal spermatogenesis in *Drosophila*. In *Biology of Drosophila*, ed. M. Demerec, pp. 1–61. New York: Hafner.

Davis, H. S. (1908). Spermatogenesis in *Acrididae* and *Locustidae*. *Bull. Mus. comp. Zool. Harvard Coll.* **53**, 59–158.

Deegener, P. (1928). Geschlechtsorgane. In *Handbuch der Entomologie*, ed. C. Schröder, vol. 1, chap. 8. Jena: G. Fischer.

Demakidoff, K. (1902). Zur Kenntnis des Baues des Insektenhodens. *Zool. Anz.* **25**, 575–8.

Demandt, C. (1912). Der Geschlechtsapparat von *Dytiscus marginalis*. *Z. wiss. Zool.* **103**, 171–299.

Depdolla, P. (1928). Die Keimzellenbildung und die Befruchtung bei den Insekten. In *Handbuch der Entomologie*, ed. C. Schröder, vol. 1, chap. 11. Jena: G. Fischer.

Dumser, J. B. & Davey, K. G. (1975). The *Rhodnius* testis: hormonal effects on germ cell division. *Can. J. Zool.* **53**, 1682–9.

Edwards, J. S. (1961). On the reproduction of *Prionoplus reticularis* (Coleoptera, Cerambycidae) with general remarks on reproduction in the Cerambycidae. *Q. J. microsc. Sci.* **102**, 519–29.

Eisentraut, M. (1926). Die spermatogonialen Teilungen bei Acridiern. *Z. Zool.* **127**, 141–284.

Fowler, G. L. (1973). *In vitro* cell differentiation in the testes of *Drosophila hydei*. *Cell Differ.* **2**, 33–42.

Friele, A. (1930). Die postembryonale Entwicklungsgeschichte der männlichen Geschlechtsorgane und Ausführungsgänge von *Psychoda alternata*. *Z. Morph. Oekol.* **18**, 249–88.

Fujimura, W., Masuda, T. & Ueyama, S. (1957). Electron microscopy of the nutritive cell of testis in the insect. *J. Nara Med. Assoc.* **8**, 177–80.

Garcia-Bellido, A. (1964*a*). Beziehungen zwischen Vermehrungswachstum und Differenzierung männlicher Keimzellen von *Drosophila melanogaster*. *Roux' Arch.* **155**, 594–610.

Garcia-Bellido, A. (1964*b*). Analyse der physiologischen Bedingungen des Vermehrungswachstums männlicher Keimzellen von *Drosophila melanogaster*. *Roux' Arch.* **155**, 611–31.

Gilmour, D. (1961). *The Biochemistry of Insects.* New York: Academic Press.

Gilson, G. (1884). I. Etude comparée de la spermatogénèse chez les Arthropodes. *Cellule* **1**, 32–7.

Goldschmidt, R. (1917). Versuche zur Spermatogenese *in vitro*. *Arch. Zellforsch.* **14**, 421–50.

Goldschmidt, R. (1931). Neue Untersuchungen über die Umwandlung der Gonaden bei intersexuellen *Limantia dispar* R. *Roux' Arch.* **124**, 618–53.

Grünberg, K. (1903). Untersuchungen über die Keim- und Nährzellen in den Hoden und Ovarien der *Lepidopteren*. *Z. wiss. Zool.* **74**, 327–95.

Hadorn, E., Remensberger, P. & Tobler, H. (1964). Autonomie in der Hodenentwicklung und Dissoziation von Chemogenese und Histogenese bei *Drosophila melanogaster*. *Rev. suisse Zool.* **71**, 583–91.

Hannah-Alava, A. (1965). The premeiotic stages of spermatogenesis. *Adv. Genet.* **13**, 157–226.

Hegner, R. (1914). Studies on germ cells. I. The history of the germ cells in insects with special reference to the Keimbahn-Determination. II. The origin and significance of the Keimbahn-Determinants in animals. *J. Morph.* **25**, 375–509.

Henderson, W. D. (1907). Zur Kenntnis der Spermatogenese von *Dytiscus marginalis* nebst Bemerkungen über den Nucleolus. *Z. wiss. Zool.* **87**, 644–84.

Hess, O. & Meyer, G. F. (1968). Genetic activities of the Y chromosome in *Drosophila* during spermatogenesis. *Adv. Genet.* **14**, 171–223.

Hirschler, J. (1953). Die Organisation des männlichen Geschlechtszellenverbandes von *Macrothylacia rubi* L. (*Lepidoptera*) (Ein methodischer Versuch). *Zool. Jahrb.* (*Anat.*) **73**, 229–75.

Hoage, T. R. & Kessel, R. G. (1968). An electron microscope study of the process of differentiation during spermatogenesis in the drone honey bee (*Apis mellifera* L.) with special reference to centriole replication and elimination. *J. Ultrastruct. Res.* **24**, 6–32.

Holmgren, H. (1901). Ueber den Hoden und die Spermatogenese von *Staphylinus*. *Anat. Anz.* **19**, 449–61.

Holmgren, H. (1902). Ueber den Hoden und die Spermatogenese von *Silpha carinata*. *Anat. Anz.* **22**, 194–206.

Hughes-Schrader, S. (1935). The chromosome cycle of *Phenacoccus* (Coccidae). *Biol. Bull.* **69**, 462.

Hughes-Schrader, S. (1946). A new type of spermiogenesis in iceryine coccids – with linear alignment of chromosomes in the sperm. *J. Morph.* **78**, 43–83.

Imms, A. D. (1957). *General Textbook of Entomology*, 9th edn. London: Methuen & Co.

Jones, J. C. (1967). Spermatocysts in *Aedes aegypti* (Linnaeus). *Biol. Bull.* **132**, 23–33.

Junker, H. (1923). Cytologische Untersuchungen an den Geschlechtsorganen der halbzwittrigen Steinfliege *Perla marginata* (Panyer). *Arch. Zellforsch.* **17**, 185–259.

Kawaguchi, E. (1928). Cytologische Untersuchungen am Seidenspinner und seiner Verwanten. I. Gametogenese von *Bombyx mori* L. und *Bombyx mandarina* M. und ihrer Bastarde. *Z. Zellforsch. mikrosk. Anat.* **7**, 519–52.

Kenchenius, P. E. (1913). The structure of the internal genitalia of some male *Diptera*. *Z. wiss. Zool.* **105**, 501–36.

Knaben, N. (1931). Spermatogenese bei *Tischeria angusticolella* Dup. *Z. Zellforsch. mikrosk. Anat.* **13**, 291–323.

Kornhauser, S. I. (1914). A comparative study of the chromosomes in the spermatogenesis of *Euchenopa binotata* Say and *Euchenopa curvata* Fabr. *Arch. Zellforsch.* **12**, 241–98.

Laugé, G. (1970). Problémes posés par les insects concernant la différenciation du sexe. *Bull. Soc. zool. Fr.* **95**, 363–77.

Lomen, F. (1914). Der Hoden von *Culex pipiens*. *Jena Z. Naturwiss.* **52**, 567–628.

Lutmann, B. F. (1910). The spermatogenesis of the caddisfly (*Platyphylax designatus* Wolk). *Biol. Bull.* **19**, 55–72.

Lyapunova, N. A. & Zoslimovskaya, A. I. (1972). Autoradiographic study of the duration of spermatogenic stages in the cricket. *Tsitologia* **15**, 276–83; *Biol. Abstr.* **57**, 294.

McClung, C. E. (1938). The apical cell of the insect testis – a possible function. *Trav. Stat. Zool. Wimereux* **13**, 437–44.

Mackinnon, E. A. & Basrur, P. K. (1970). Cytokinesis in the gonocysts of the drone honey bee (*Apis mellifera* L.) *Can. J. Zool.* **48**, 1163–6.

Masner, P. (1965). The structure, function and imaginal development of the male inner reproductive organs of *Adelphocoris lineolatus* (Goeze) (*Heteroptera*, Miridae). *Acta Ent. Bohemoslov.* **62**, 254–76.

Menon, M. (1969). Structure of the apical cells of the testis of the Tenebrionid beetles, *Tenebrio molitor* and *Zaphobas rugipes*. *J. Morph.* **127**, 409–30.

Metz, C. W. (1938). Chromosome behavior inheritance and sex determination in *Sciara*. *Am. Nat.* **72**, 485–520.

Metz, C. W. & Nonidez, F. (1921). Spermatogenesis in the fly *Asilus sericeus* Say. *J. exp. Zool.* **32**, 165–86.

Meves, F. (1907). Die Spermatozytenteilungen bei der Honigbiene (*Apis mellifica* L.), nebst einigen Bemerkungen über Chromatinreduktion. *Arch. mikrosk. Anat.* **70**, 414–91.

Meyer, G. F. (1961). Interzelluläre Brücken (Fusome) im Hoden und im Ei-Nährzellenverband von *Drosophila melanogaster*. *Z. Zellforsch. mikrosk. Anat.* **54**, 238–51.

Mohr, O. L. (1914). Studien über die Chromatinreifung der männlichen Geschlechtszellen bei *Locusta viridissima*. *Arch. Biol.* **29**, 579–752.

Montgomery, T. H. (1910). On the dimegalous sperm and chromosomal variation of *Euschistus* with reference to chromosomal continuity. *Arch. Zellforsch.* **5**, 120–45.

Muckenthaler, F. A. (1964). Autoradiographic study of nucleic acid synthesis during spermatogenesis in the grasshopper *Melanoplus differentialis*. *Exp. Cell Res.* **35**, 531–97.

Naisse, J. (1970). Influence des hormones sur la différenciation sexuelle de *Lampyris noctiluca* (Coléoptère). *Bull. Soc. zool. Fr.* **95**, 377–82.

Nelsen, O. E. (1931). Life cycle, sex differentiation, and testis development in *Melanoplus differentialis* (Acrididae, Orthoptera). *J. Morph.* **51**, 467–525.

Nur, U. (1962). Sperms, sperm bundles and fertilization in a mealy bug *Pseudococcus obscurus* Essig. (*Homoptera*: Coccidae). *J. Morph.* **111**, 173.

Perrot, J. L. (1933). Le spermatogénèse et l'ovogénèse de *Lepisma* (*Thormolia*) *domestica*. Heteropycnose dans un sexe homogamétique. *Z. Zellforsch. mikrosk. Anat.* **18**, 573–92.

Perrot, J. L. (1934). La spermatogénèse et l'ovogénèse du Mallophage *Goniodes stylifer*. *Q. J. microsc. Sci.* **76**, 353–77.

Phillips, D. M. (1970). Insect sperm: their structure and morphogenesis. *J. Cell Biol.* **44**, 243–77.

Poisson, R. (1936). Nouvelles observations sur les processes spermatogénètiques sur les éléments sexuels d'hemipteres aquatiques (*Velia currens* Fab. et Gerris). *Arch. Zool. exp. gén.* **78**, 133–94.

Schellenberg, A. (1913). Das accessorische Chromosom in den Samenzellen der Locustide *Diestrammena marmorata* de Hahn. *Arch. Zellforsch.* **11**, 489–514.

Schneider, K. (1917). Die Entwicklung des Eierstockes und Eies von *Deilephila euphorbiae.* *Arch. Zellforsch.* **14**, 79–143.

Schrader, F. (1945a). The cytology of regular heteroploidy in the genus *Loxa* (Pentatomidae – Hemiptera). *J. Morph.* **76**, 157–77.

Schrader, F. (1945b). Regular occurrence of heteroploidy in a group of *Pentatomidae* (Hemiptera). *Biol. Bull.* **88**, 63–70.

Smith, D. S. (1968). *Insect Cells: Their Structure and Function.* Edinburgh: Oliver & Boyd.

Snodgrass, R. E. (1935). *Principles of Insect Morphology.* New York: McGraw-Hill.

Spichardt, C. (1886). Beitrag zur Entwicklung der männlichen Genitalien und ihrer Ausführungsgänge bei *Lepidopteren. Verh. Naturhist. Verein Rheinland, Bonn* **43**, 1–43.

Szöllösi, A. (1975). Electron microscope study of spermiogenesis in *Locusta migratoria* (Insecta Orthoptera). *J. Ultrastruct. Res.* **50**, 322–46.

Tannreuther, G. W. (1907). History of the germ cells and early embryology of certain aphids. *Zool. Jahrb. (Anat.).* **24**, 609–42.

Tokuyasu, K. T., Peacock, W. J. & Hardy, R. W. (1972). Dynamics of spermiogenesis in *Drosophila melanogaster.* I. Individualization process. *Z. Zellforsch. mikrosk. Anat.* **124**, 479–506.

Verson, E. (1889). Zur Spermatogenesis. *Zool. Anz.* **12**, 100–3.

Wake, K. (1963). Cytological studies on nuclear division in the wall of the testis of an Orthoptera, *Acrida lata* Motschulsky. *Osaka City Med. J.* **9**, 129–52.

White, M. J. D. (1946). The cytology of the *Cecidomyidae* (*Diptera*). II. The chromosome cycle and anomalous spermatogenesis of *Miastor. J. Morph.* **79**, 323–70.

White, M. J. D. (1947). The cytology of the *Cecidomyidae* (*Diptera*). III. The spermatogenesis of *Taxomia taxi. J. Morph.* **80**, 1–24.

White, M. J. D. (1955). Patterns of spermatogenesis in grasshoppers. *Aust. J. Zool.* **3**, 222–6.

Wiemann, H. L. (1910a). A study in the germ cells of *Leptinotarsa signaticollis. J. Morph.* **21**, 135–216.

Wiemann, H. L. (1910b). The degenerated cells in the testis of *Leptinotarsa signaticollis.* *J. Morph.* **21**, 485–94.

Wittig, G. (1960). Morphologie und Entwicklung der Raupen des Tannentriebwicklers *Choristoneura marinana* (HB.) Lepidopt. Tortricidae. II. Die Entwicklung der Geschlechtsorgane. *Zool. Jahrb. (Anat.),* **78**, 145–66.

Zick, K. (1911). Die postembryonale Entwicklung des männlichen Geschlechtsapparats der Schmetterlinge. *Z. wiss. Zool.* **98**, 430–77.

Zweiger, H. (1906). Die Spermatogenese von *Forficula auricularia. Zool. Anz.* **30**, 220–6.

Chapter 7. Various invertebrate phyla

Anderson, W. A., Weisman, A. & Ellis, R. A. (1967). Cytodifferentiation during spermiogenesis in *Lumbricus terrestris. J. Cell Biol.* **32**, 11–26.

Atwood, D. G. (1974). Fine structure of spermatogonia, spermatocytes, and spermatids of the sea cucumbers *Cucumaria lubrica* and *Leptosynapta clarki* (Echinodermata: Holothuroidea). *Can. J. Zool.* **52**, 1389–96.

Austin, C. R. (1964). Gametogenesis and fertilization in the mesozoan *Dicyema aegira. Parasitology* **54**, 597–600.

Beams, H. W. & Sekhon, S. S. (1972). Cytodifferentiation during spermiogenesis in *Rhabditis pellio*. *J. Ultrastruct. Res.* **38**, 511–27.

Beddard, F. E. (1895). *A Monograph on the Order of Oligochaeta*. Oxford: Clarendon Press.

Bloomfield, J. E. (1880). On the development of spermatozoa. Part I. *Lumbricus. Q. J. microsc. Sci.* **20**, 79–89.

Bondi, C. & Farnesi, R. M. (1976). EM studies of spermatogenesis in *Branchiobdella pentadonta* Whitman (Annelida, Oligochaeta). *J. Morph.* **148**, 65–88.

Brien, P. (1960). Classe de Bryozosires. In *Traité de Zoologie*, ed. P. P. Grassé, vol. 5, pp. 1143–4. Paris: Masson et Cie.

Brien, P. (1966). *Biologie de la Reproduction Animale*. Paris: Masson et Cie.

Buchner, P. (1914). Die Besamung der jugendlichen Ovocyte und die Befruchtung des Saccocirrus. *Arch. Zellforsch.* **12**, 395–414.

Bugnion, E. & Popoff, N. (1905). La spermatogénèse du lombric terrestre. *Arch. Zool. exp. gén.* **3**, 339–89.

Bunke, D. (1967). Zur Morphologie und Systematik der Aeolosomatidae Beddard 1895 und Potamodrilidae nov. fam. (Oligochaeta). *Zool. Jahrb.* (*Syst.*) **94**, 187–368.

Burton, P. R. (1960). Gametogenesis and fertilization in the frog lung fluke, *Haematoloechus medioplexus* (Trematoda: Plagiorchiidae). *J. Morph.* **103**, 93–122.

Burton, P. R. (1972). Fine structure of the reproductive system of a frog lung fluke. III. The spermatozoan and its differentiation. *J. Parasit.* **58**, 68–83.

Caullery, M. (1911). The gonads of the urchin *Echinocardium cordatum*. *Nature, Lond.* **88**, 26.

Chatton, E. & Tuzet, O. (1941). Sur quelques faits nouveaux de la spermiogénèse de *Lumbricus terrestris*. *C. R. Acad. Sci., Paris* **213**, 373–6.

Chen, Pin-Dji (1937). Germ cell cycle in *Paragonimus hellicotti* Ward. *Trans. Am. microsc. Soc.* **56**, 208–36.

Child, C. M. (1907). Studies on the relation between amitosis and mitosis. II. Development of the testes and spermatogenesis in *Moniezia*. *Biol. Bull.* **12**, 175–224.

Chitwood, B. G. & Chitwood, M. B. (1940). The reproductive system. In *An Introduction to Nematology*, chap. 10. Baltimore: Monumental Printing.

Cobb, N. A. (1925). Nemic spermatogenesis. *J. Hered.* **16**, 357–9.

Cognetti, G. & Delavault, R. (1958). Aspetti istologici delle gonadi in *Coscinasterias tennispina* Lmk. *Monit. zool. ital.* **66**, 44–8.

Daly, J. M. (1974). Gametogenesis in *Harmothoe imbricata*. *Mar. Biol.* **25**, 35–40.

Das, N. K. (1968). Developmental features and synthetic patterns of male germ cells in *Urechis caupo*. *Roux's Arch.* **171**, 325–35.

Das, N. K. & Alfert, M. (1968). Cytochemical studies on the concurrent synthesis of DNA and histone in primary spermatocytes of *Urechis caupo*. *Exp. Cell Res.* **49**, 51–8.

Depdolla, P. (1906). Beiträge zur Kenntnis der Spermatogenese beim Regenwurm. *Z. wiss. Zool.* **81**, 632–90.

Dingler, M. (1910). Über die Spermatogenese des *Dicrocoelium lanceolatum*. *Arch. Zellforsch.* **4**, 672–712.

Dixon, G. C. (1915). *Tubifex. Liverpool Marine Biology Communication, Memoir 23*. London.

Fauré-Fremiet, E. (1913). Le cycle germinatif chez l'*Ascaris megalocephala*. *Arch. Anat. microsc.* **15**, 435–757.

Favard, P. (1951). The evolution of the ergastoplasm in spermatids of *Ascaris*. *C. R. Acad. Sci., Paris* **248**, 3344–6.

Fedecka-Bruner, B. (1965). Régénération des testicules des planaires après destruction par les rayons X. In *Regeneration in Animals and Related Problems: an International Symposium*, ed. V. Kiortsis & H. A. L. Trampusch, pp. 185–92. Amsterdam: North-Holland Publishing Co.

Franzén, Z. (1973). The spermatozoon of *Siboglinum* (Pogonophora). *Act. Zool.* **54**, 179–92.

Frauquinnet, R. & Lender, T. (1973). Etude ultrastructurale des testicules de *Polycelis tenius* et *Polycelis nigra* (Planaires). Evolution des cellules germinales mâles avant la spermatogénèse. *Z. mikrosk.-anat. Forsch.* **87**, 2–44.

Ghiradelli, E. (1965). Differentiation of the germ cells and regeneration of the gonads in Planarians. In *Regeneration in Animals and Related Problems: an International Symposium*, ed. V. Kiortsis & H. A. L. Trampusch, pp. 177–84. Amsterdam: North-Holland Publishing Co.

Grellet, P. (1957). Histologie du testicule et spermatogénèse chez *Alcyonidium gelatinosum* (L.) (Bryozoaire eténostome). *Bull. Lab. mar. Dinard* **43**, 3–24.

Guilford, H. G. (1955). Gametogenesis in *Heronimus chelydrae* MacCallum. *Trans. Am. microsc. Soc.* **74**, 182–90.

Henley, C. (1974). Platyhelminthes (Turbellaria). In *Reproduction of Marine Invertebrates*, ed. A. C. Giese & J. S. Pearse, chap. 5. New York: Academic Press.

Hertwig, O. (1880). Die Chaethognathen. *Jena Z. Naturwiss.* **14**, 196–311.

Hertwig, O. (1890). Vergleich der Ei- und Samenbildung bei den Nematoden. *Arch. mikrosk. Anat.* **36**, 1–138.

Holland, N. D. (1967). Gametogenesis during the annual reproductive cycle in a cidaroid sea urchin (*Stylocidaris affinis*). *Biol. Bull.* **133**, 578–90.

Holland, N. D. & Giese, A. C. (1965). An autoradiographic investigation of the gonads of the Purple Sea Urchin (*Strongylocentrotus purpuratus*). *Biol. Bull.* **128**, 241–58.

Hope, W. D. (1974). Nematoda. In *Reproduction in Marine Invertebrates*, ed. A. C. Giese & J. S. Pearse, chap. 8, pp. 391–469. New York: Academic Press.

Humes, A. G. (1941). The male reproductive system in the nemertean genus *Carcinonemertes*. *J. Morph.* **69**, 443–54.

Hyman, L. (1951). *The Invertebrates: Platyhelminthes and Rhynchocoela*, vol. 2. New York: McGraw-Hill.

Hyman, L. (1955). *The Invertebrates: Echinodermata*, vol. 4. New York: McGraw-Hill.

Hyman, L. (1959). *The Invertebrates: Small Coelomate Groups*, vol. 5. New York: McGraw-Hill.

Ikeda, I. (1903). Development of sexual organs and their products in *Phoronis*. *Annot. zool. Japan* **4**, 141–53.

Ivanov, A. V. (1963). *Pogonophora*. New York: Academic Press.

Jägersten, G. (1934). Studien über den histologischen Bau der männlichen Geschlechtsorgane und die Ausbildung des Spermiums bei Myzostomum. *Zool. Bidr. Uppsala* **15**, 1–22.

John, B. (1953). The behavior of the nucleus during spermatogenesis in *Fasciola hepatica*. *Q. J. microsc. Sci.* **94**, 41–55.

Koehler, J. K. (1965). An electron microscopic study of the dimorphic spermatozoa of *Asplanchna* (Rotifera). I. The adult testis. *Z. Zellforsch. mikrosk. Sci.* **67**, 57–76.

Koehler, J. K. & Birky, C. W. (1966). An electron microscopic study of the dimorphic spermatozoa of *Asplanchna* (Rotifera). II. The development of 'atypical spermatozoa'. *Z. Zellforsch. mikrosk. Anat.* **70**, 303–21.

Kozloff, E. N. (1969). Morphology of the orthonectid *Rhopalura ophiocomae*. *J. Parasit.* **55**, 171–95.

Lee, A. B. (1887). La spermatogénèse chez les chaetognathes. *Cellule* **4**, 107–33.

Lentati, G. B. (1970). Gametogenesis and egg fertilization in Planarians. *Int. Rev. Cytol.* **27**, 101–79.

Lindner, E. (1914). Über die Spermatogenese vom *Schistosomum haematobium* Bilh. (*Bilharzia haematobia* Cobb) mit besonderer Berücksichtigung der Geschlechtschromosomen. *Arch. Zellforsch.* **12**, 516–38.

Longo, F. J. & Anderson, E. (1969). Sperm differentiation in the sea urchin *Arbacia punctulata* and *Strongylocentrotus purpuratus*. *J. Ultrastruct. Res.* **27**, 486–509.

Marcus, E. (1934). Über *Lophopus crystallinus. Zool. Jahrb. (Anat.)* **58**, 501–606.

Nagano, T. (1969). The crystalloid of Lubarsch in the human spermatogonium. *Z. Zellforsch. mikrosk. Anat.* **97**, 491–501.

Newby, W. W. (1940). The embryology of the echiuroid worm *Urechis caupo. Mem. Am. Phil. Soc.* **16**, 1–219.

Nez, M. M. (1954). Gametogenesis in *Schistosomatium douthitti* (Cort.) *J. Parasit.* **40** (Suppl.), 37 (abstract).

Nigon, V. & Delavault, R. (1952). L'évolution des acides nucleique dans les cellules reproductives d'un Nematode pseudogame. *Arch. Biol.* **63**, 393–410.

Nouvel, H. (1947). Les Dicyémides. I. Systématique, générations vermiformes, Infusorigène et Sexualité. *Arch. Biol.* **58**, 59–220.

Olive, P. J. W. (1972*a*). Regulation and kinetics of spermatogonial proliferation in *Arenicola marina* (Annelida: Polychaeta). I. The annual cycle of mitotic index in the testis. *Cell Tiss. Kinet.* **5**, 245–53.

Olive, P. J. W. (1972*b*). Regulation and kinetics of spermatogonial proliferation in *Arenicola marina* (Annelida: Polychaeta). II. Kinetics. *Cell Tiss. Kinet.* **5**, 255–67.

Pasteels, J. (1951). Recherches sur le cycle germinal de l'*Ascaris. Arch. Biol.* **59**, 405–46.

Patent, D. H. (1969). The reproductive cycle of *Gorgonocephalus caryi* (Echinodermata: Ophiuroidea). *Biol. Bull.* **136**, 241–52.

Prestage, J. J. (1960). The fine structure of the growth region of ovary in *Ascaris lumbricoides* (var. Suum) with special reference to the rachis. *J. Parasit.* **46**, 69–78.

Rappeport, T. (1917). Zur Spermatogenese der Süsswasser-Tricladen. *Arch. Zellforsch.* **14**, 1–25.

Rattenburg, J. (1953). Reproduction in *Phoronopsis viridis. Biol. Bull.* **104**, 182–96.

Rice, M. E. (1974). Gametogenesis in three species of Sipuncula: *Phascolosoma agassizii, Golfingia pugettensis,* and *Themiste pyroides. Cellule* **70** ", 295–313.

Ridley, R. K. (1969). Electron microscopic studies on dicyemid. II. Infusorigen and infusoriform stages. *J. Parasit.* **55**, 779–93.

Riepen, O. (1933). Anatomie und Histologie von *Malacobdella grossa* (Müll.) *Z. wiss. Zool.* **143**, 323–496.

Riser, N. W. (1974). Nemertinea. In *Reproduction of Marine Invertebrates,* ed. A. C. Giese & J. S. Pearse, vol. 1, chap. 7. New York: Academic Press.

Rosario, B. (1964). An electron microscope study of spermatogenesis in Cestodes. *J. Ultrastruct. Res.* **11**, 412–27.

Rothschild, L. (1965). *A Classification of Living Animals.* London: Longmans, Green & Co.

Salensky, W. (1907). Morphogenetische Studien an Würmern. II. Ueber den Bau der Archianneliden nebst Bemerkungen über den Bau einiger Organe des *Saccocirrus papillocercus* Bobr. *Mem. Acad. Imp. Sci., Petersburg* **16**, 103–451.

Sato, M., Oh, N. & Sakoda, K. (1967). EM study of spermatogenesis in the lung fluke (*Paragonimus Miyazakii*). *Z. Zellforsch. mikrosk. Anat.* **77**, 232–43.

Schleip, W. (1907). Die Samenreifung bei den Planarien. *Zool. Jahrb. (Anat.)* **24**, 129–74.

Severinghaus, A. E. (1927). Sex studies on *Schistosoma japonicum. Q. J. microsc. Sci.* **71**, 653–702.

Silen, L. (1966). On the fertilization problem in gymnolaematous Bryozoa. *Ophelia* **3**, 113–40.

Stang-Voss, C. (1970). Zur Entstehung des Golgi-Apparates. Elektronenmikroskopische Untersuchungen an Spermatiden von *Eisenia foetida* (Annelidae). *Z. Zellforsch. mikrosk. Anat.* **109**, 287–96.

Stephenson, J. (1922). Contributions to the morphology, classification and zoogeography of Indian Oligochaeta. IV. On the diffuse production of sexual cells in a species of *Chaetogaster* (fam. Neididae). *Proc. Zool. Soc., London* 109–48.

Stephenson, J. (1930). *The Oligochaeta.* Oxford: Clarendon Press.

Tanaka, Y. (1958). Seasonal changes occurring in the gonad of *Stichopus japonicus. Bull. Fac. Fish. Hokkaido Univ.* **9**, 29–36.

Tangapregassom, A. M. & Delavault, R. (1967). Analyse, en microscopie photonique et électronique, des structures périphériques des gonades chez deux étioles de mer: *Asterina gibbosa* Peunant et *Echinaster sepositus* Gray. *Cah. Biol. mar.* **8**, 153–9.

Tauson, A. (1926). Die Spermatogenese bei *Asplanchna intermedia* Huds. *Z. Zellforsch. mikrosk. Anat.* **4**, 652–81.

Taylor, A. E. R. (1960). The spermatogenesis and embryology of *Litomosoides carinii* and *Dirofilaria immitis. J. Helminth.* **34**, 3–12.

Triantophyllou, A. C. & Hirschmann, H. (1964). Reproduction in plant and soil nematodes. *Am. Rev. Phytopath.* **2**, 57–80.

Tuzet, O. (1945). Sur la Spermatogénèse atypique lombriciens. *Arch. Zool. exp. gén.* **84**, 155–68.

Van Beneden, E. & Julin, C. (1884). La spermatogénèse chez l'*Ascaris mégalocephale. Bull. Acad. R. Belg.* **7**, 312–42.

Van Der Woude, A. (1954). Germ cell cycle of *Megalodiscus temperatus* (Stafford, 1905) Harwood, 1932 (Paramphistomidae, Trematoda). *Am. Midl. Nat.* **51**, 172–202.

Voigt, W. (1885). Ueber Ei- und Samenbildung bei *Branchiobdella. Arb. Zool. Zootom. Hist.,* Würzburg **7**, 300–67.

Walker, C. W. (1974). Studies on the reproductive systems of sea-stars. I. The morphology and histology of the gonad of *Asterias vulgaris. Biol. Bull.* **147**, 661–77.

Walsh, M. P. (1954). Spermatogenesis of *Lumbricus terrestris* (L.). *Trans. Am. microsc. Soc.* **73**, 59–65.

Walton, A. C. (1924). Studies on nematode gametogenesis. *Z. Zellforsch. mikrosk. Anat.* **1**, 167–239.

Walton, A. C. (1940). Gametogenesis. In *An Introduction into Nematology, Section II,* ed. B. G. Chitwood & M. B. Chitwood, chap. 1. Baltimore: Monumental Printing.

Weygandt, C. (1907). Beiträge zur Kenntnis der Spermatogenese bei *Plagiostomum Girardi. Z. wiss. Zool.* **88**, 249–90.

Whitman, C. O. (1883). A contribution to the embryology, life history, and classification of the dicyemids. *Mitt. Zool. Stn., Neapel* **4**, 1–89.

Willey, C. H. & Goodman, G. C. (1951). Gametogenesis, fertilization and cleavage in the trematode *Zygocotyle lunata* (Paramphistomidae). *J. Parasit.* **37**, 283–96.

Willey, C. H. & Koulish, S. (1950). Development of germ cells in the adult stage of the digenetic trematode, *Gorgoderina attenuta* Stafford, 1902. *J. Parasit.* **36**, 67–79.

Willmer, E. N. (1970). *Cytology and Evolution.* New York: Academic Press.

Willmott, S. (1950). Gametogenesis and early development in *Gigantocotyle bathycotyle* (Fischoeder, 1901) Näsmark, 1937. *J. Helminth.* **24**, 1–14.

Wilson, L. P. (1940). Histology of the gonad wall of *Arbacia punctulata. J. Morph.* **66**, 463–79.

Woodhead, A. E. (1931). Germ cell cycle in the trematode family *Bucephalidae. Trans. Am. microsc. Soc.* **50**, 169–88.

Woodhead, A. E. (1957). Germ cell development in the first and second generations of *Schistosomatium Douthitti* (Cort, 1914) Price, 1931 (Trematoda, Schistosomatidae). *Trans. Am. microsc. Soc.* **76**, 173–6.

Whitney, D. C. (1917). The production of functional and rudimentary spermatozoa in rotifers. *Biol. Bull.* **33**, 305–15.

Wilson, E. B. (1925). *The cell in Development and Heredity.* New York: Macmillan.

Yosufzai, H. K. (1952). Cytological studies on the spermatogenesis of *Fasciola hepatica* L. *Cellule* **55**, 7–18.

Young, R. T. (1923). Gametogenesis in Cestodes. *Arch. Zellforsch.* **17**, 419–37.

Chapter 8. Vertebrata

The references for this chapter are split into three sections covering fish, Amphibia and Amniota (Reptilia, Aves and Mammalia) respectively.

Fish

Ahsan, S. N. (1966a). Cyclical changes in the testicular activity of the lake chub, *Conesius plumbeus* (Agassiz). *Can. J. Zool.* **44**, 149–59.

Ahsan, S. N. (1966b). Effects of temperature and light on the cyclical changes in the spermatogenetic activity of the lake chub, *Conesius plumbeus* (Agassiz). *Can. J. Zool.* **44**, 161–71.

Allen, E. (1918). Studies on cell division in the albino rat. III. Spermatogenesis: the origin of the first spermatocytes and the organization of the chromosomes, including the accessory. *J. Morph.* **31**, 133–74.

Bennington, N. L. (1936). Germ cell origin and spermatogenesis in the Siamese fighting fish, *Betta splendens*. *J. Morph.* **60**, 103–25.

Billard, R. (1968). Influence de la température sur la durée et l'efficacité de la spermatogenèse du Guppy *Poecilia reticulata*. *C. R. Acad. Sci., Paris* **266**, 2287–90.

Billard, R. (1969a). La spermatogénèse de *Poecilia reticulata*. I. Estimation du nombre de générations goniales et rendement de la spermatogenèse. *Ann. Biol. anim. Biochim. Biophys.* **9**, 251–71.

Billard, R. (1969b). La spermatogenèse de *Poecilia reticulata*. II. La production spermatogénétique. *Ann. Biol. anim. Biochim. Biophys.* **9**, 272–80.

Billard, R. (1970). La spermatogenèse de *Poecilia reticulata*. III. Ultrastructure des cellules de Sertoli. *Ann. Biol. anim. Biochim. Biophys.* **10**, 37–50.

Billard, R., Breton, B. & Jalabert, B. (1970). La production spermatogénétique chez la truite. *Ann. Biol. anim. Biochim. Biophys.* **11**, 190–212.

Billard, R. & Fléchon, J. E. (1969). Spermatogonies et spermatocytes flagellés chez *Poecilia reticulata* (Téléostéens cyprinodontiformes). *Ann. Biol. anim. Biochim. Biophys.* **9**, 281–6.

Billard, R., Jalabert, B. & Breton, B. (1972). Les cellules de Sertoli des poissons téléostéens. *Ann. Biol. anim. Biochim. Biophys.* **12**, 19–32.

Bowers, A. B. & Holliday, F. G. T. (1969). Histological changes in the gonads associated with the reproductive cycle of the herring (*Clupea harengus* L.). *Mar. Res. Ser. Scottish Home Dept.* **5**, 1–16.

Brock, J. (1878). Beiträge zur Anatomie und Histologie der Geschlechtsorgane der Knochenfische. *Morph. Jahrb.* **4**, 505–72.

Bullough, W. S. (1939). A study of the reproductive cycle of the minnow (*Phoxinus laevis*, L.) in relation to the environment. *Proc. Zool. Soc. Lond., Ser. A* **109**, 79–102.

Butcher, E. O. (1929). The origin of the germ cells in the lake lamprey (*Petromyzon murinus unicolor*). *Biol. Bull.* **56**, 87–98.

Champy, C. (1921). La structure remarquable de testicule des blennies. *C. R. Assoc. Anat., Paris* **16**, 165–70.

Chan, S. T. H. & Phillips, J. G. (1967). The structure of the gonad during natural sex reversal in *Monopterus albus* (Pisces: Teleostei). *J. Zool., Lond.* **151**, 129–41.

Chavin, W. & Gordon, M. (1951). Sex determination in *Platypoecilus maculatus*. I. Differentiation of the gonads in members of all-male broods. *Zoologica* **36**, 135–45.

Clérot, J. C. (1971). Les ponts intercellulaires du testicule du Gardon: organisation syncitiale et synchronie de la différenciation des cellules germinales. *J. Ultrastruct. Res.* **37**, 690–703.

Craig-Bennet, A. (1931). The reproductive cycle of the three-spined stickleback (*Gasterosteus aculeatus*, Linn.). *Phil. Trans. R. Soc. Lond., Ser. B* **291**, 197–280.

Cunningham, J. T. (1886). On the structure and development of the reproductive elements in *Myxine glutinosa* L. *Q. J. microsc. Sci.* **27**, 49–76.

D'Ancona, V. (1943). Nuove recirche sulla determinazion sessuale dell'anguilla. *Arch. Oceanogr. Limnol., Roma* **3**, 159–271.

De Felice, D. A. & Rasch, E. M. (1969). Chronology of spermatogenesis and spermiogenesis in *Poeciliid* fishes. *J. exp. Zool.* **171**, 191–208.

Dildine, G. C. (1936). Studies in teleostean reproduction. I. Embryonic hermaphroditism in *Lebistes reticulatus. J. Morph.* **60**, 261–78.

Dodds, G. S. (1910). Segregation of the germ cells of the teleost, *Lophius. J. Morph.* **21**, 563–611.

Dulzetto, F. (1933). La struttura del testicola di *Gambusia holbrookii* (Grd.) *Arch. Zool., Napoli* **19**, 405–37.

Egami, N. & Hyodo-Taguchi, V. (1967). An autoradiographic examination of the rate of spermatogenesis at different temperatures in the fish, *Oryzias latipes. Exp. Cell Res.* **47**, 665–7.

Ewing, H. (1972). Spermatogenesis in the zebra fish *Brachydanio rerio* (Hamilton-Buchanan). *Anat. Rec.* **172**, 308.

Foley, J. D. (1926). The spermatogenesis of *Umbra limi* with special reference to the behaviour of the spermatogonial chromosomes and the first maturation division. *Biol. Bull.* **50**, 117–46.

Fontaine, M. & Tuzet, O. (1937). Sur la spermatogénèse de l'anguille argentée. *Arch. zool. exp.* **78**, 199–215.

Geiser, S. W. (1924). Sex ratios and spermatogenesis in the top-minnow, *Gambusia holbrookii* Grd. *Biol. Bull.* **47**, 175–208.

Gokhale, S. V. (1957). Seasonal histological changes in the gonads of the whiting (*Gadus merlangus* L.) and the Norway pout (*Gadus esmarkii* Nilsoon). *Indian J. Fish.* **4**, 92–112.

Goodrich, H. B., Dee, J. E., Flynn, C. M. & Mercer, R. M. (1934). Germ cells and sex differentiation in *Lebistes reticulatus. Biol. Bull.* **67**, 83–96.

Grier, H. J. (1975). Aspects of germinal cyst and sperm development in *Poecilia latipinna* (Teleostei: Poeciliidae). *J. Morph.* **146**, 229–50.

Hann, H. W. (1927). The histology of the germ cells of *Cottus bairdii* Girard. *J. Morph.* **43**, 427–98.

Hardisty, M. W. (1965*a*). Sex differentiation and gonadogenesis in lampreys. I. The ammocoete gonads of the brook lamprey, *Lampetra planeri. J. Zool.* **146**, 305–45.

Hardisty, M. W. (1965*b*). Sex differentiation and gonadogenesis in lampreys. II. The ammocoete gonads of the landlocked sea lamprey, *Petromyzon marinus. J. Zool.* **146**, 346–87.

Henderson, N. E. (1962). The annual cycle of the testis of the eastern brook trout, *Salvelinus fontinalis* (Mitchill). *Can. J. Zool.* **40**, 631–41.

Hoffman, R. (1963). Gonads, spermatic ducts and spermatogenesis in the reproductive system of male toadfish, *Opsanus tau. Cheasapeake Sci.* **4**, 21–9.

Holstein, A. F. (1969). Zur Frage der lokalen Steuerung der Spermatogenese beim Dornhai (*Squalus acanthias* L.). *Z. Zellforsch. mikrosk. Anat.* **93**, 265–81.

Jensen, O. S. (1883). Recherches sur la spermatogénèse. *Arch. Biol., Paris* **4**, 669–747.

Johnston, P. M. (1951). The embryonic history of the germ cells of the large-mouth black bass, *Micropterus salmoides salmoides* (Lacépède). *J. Morph.* **88**, 471–542.

Jones, J. W. (1940). Histological changes in the testes in the sexual cycle of the male salmon parr (*Salmo salar* L. Guv.). *Proc. Roy. Soc. Lond. Ser. B* **128**, 499–509.

Lofts, B. (1968). Patterns of testicular activity. In *Perspectives in Endocrinology*, ed. E. J. W. Barrington & C. B. Jørgensen, chap. 4, pp. 239–304. New York: Academic Press.

Lofts, B. & Marshall, A. J. (1957). Cyclical changes of the testis lipids of a teleost fish, *Esox lucius. Q. J. microsc. Sci.* **98**, 79–88.

Marshall, A. J. & Lofts, B. (1956). The Leydig cell homologue in certain teleost fishes. *Nature, Lond.* **177**, 705–6.

Matthews, L. H. (1950). Reproduction in the basking shark *Cetorhinus maximus* (Gunner). *Phil. Trans. R. Soc. Lond., Ser. B* **234**, 247–316.

Matthews, S. A. (1938). The seasonal cycle in the gonads of *Fundulus. Biol. Bull.* **75**, 66–74.

Mizue, K. (1958). Studies on a scorpaenous fish *Sebasticus marmoratus* Cuv. et Val. II. The seasonal cycle of mature testis and the spermatogenesis. *Bull. Fac. Fish. Nagasaki Univ.* **6**, 27–38.

Moser, H. (1967). Seasonal histological changes in the gonads of *Sebastodes paucispinis* Ayres, an ovoviviparous teleost (Family Scorpaenidae). *J. Morph.* **23**, 329–54.

Okkelberg, P. (1921). The early history of the germ cells in the brook lamprey *Entosphenus wilderi* (Gage), up to and including the period of sexual differentiation. *J. Morph.* **35**, 1–151.

Oordt, G. J. van (1929). Zur mikroskopischen Anatomie des Ovariotestis von *Serranus* und *Sargus* (Teleostei). *Z. mikrosk.-anat. Forsch.* **19**, 1–17.

Phillippi, E. (1908). Fortpflanzungsgeschichte der viviparen Teleosteen *Glaridichtys januarius* und *Glaridichtys decemmaculatus* in ihrem Einfluß auf Lebensweise, makroskopische und mikroskopische Anatomie. *Zool. Jahrb.* **27**, 1–94.

Polder, J. J. W. (1971). On gonads and reproductive behaviour in the cichlid fish *Aequidens portaligrensis* (Hensel). *Neth. J. Zool.* **23**, 265–365.

Raizada, A. K. (1975). The testicular cycle of a Percoid teleost *Nandus nandus* (Ham.). *Gegenbaurs morph. Jahrb.* **121**, 77–87.

Regaud, C. (1901). Etudes sur la structure des tubes séminifères et sur la spermatogénèse chez les Mammifères. *Arch. Anat. microsc.* **4**, 101–56, 231–80.

Rodolico, A. (1933). Differenciamento dei sessi ed ovospermatogenesi nell'anguilla. *Pubbl. Stn. zool., Napoli* **13**, 180–278.

Ruby, S. M. & McMillan, D. B. (1975). The interstitial origin of germinal cells in the testis of the stickleback *Culaea inconstans. J. Morph.* **145**, 295–318.

Shrivastava, S. S. (1967). Histomorphology and seasonal cycle of the spermary and spermduct in a teleost *Notopterus notopterus* Pollas. *Acta anat.* **66**, 133–60.

Simpson, T. H., Wright, R. S. & Hunt, S. V. (1964). Steroid biosynthesis in the testis of the dogfish *Squalus acanthias. J. Endocr.* **31**, 29–38.

Stanley, H. P. (1963). Urogenital morphology in the chimaeroid fish *Hydrolagus colliei* (Lay & Bennett). *J. Morph.* **112**, 99–127.

Stanley, H. P. (1966). The structure and development of the seminiferous follicle in *Scyliorhinus caniculus* and *Torpedo marmorata* (Elasmobranchii). *Z. Zellforsch. mikrosk. Anat.* **75**, 453–68.

Stanley, H. P. (1969). An electron microscope study of spermiogenesis in the teleost fish *Oligocottus maculosus. J. Ultrastruct. Res.* **27**, 230–43.

Stanley, H. P., Chieffi, G. & Botte, V. (1965). Histological and histochemical observations on the testis of *Gobius paganellus. Z. Zellforsch. mikrosk. Anat.* **65**, 350–62.

Swaen, A. & Masquelin, H. (1883). Etude sur la spermatogénèse. *Arch. Biol., Paris* **4**, 749– 801.

Turner, C. L. (1919). The seasonal cycle in the spermary of the perch. *J. Morph.* **32**, 681–711.

Turner, C. L. (1938). The reproductive cycle of *Brachyraphis episcopi*, an ovoviviparous Poeciliid fish, in the natural tropical habitat. *Biol. Bull.* **75**, 56–65.

Vallette St George, von la (1878). Über die Genese der Samenkörper. V. Die Spermatogenese bei den Säugethieren und beim Menschen. *Arch. mikrosk. Anat.* **15**, 261–314.

Vaupel, J. (1929). The spermatogenesis of *Lebistes reticulatus. J. Morph.* **47**, 555–87.

Weisel, G. F. (1943). A histological study of the testes of the sockeye salmon *Oncorhynchus nerka. J. Morph.* **73**, 207–29.

Wiebe, J. P. (1968). The reproductive cycle of the viviparous seaperch, *Cymatogaster aggregata* Gibbons. *Can. J. Zool.* **46**, 1221–34.

Wolf, L. E. (1931). History of the germ cells in the viviparous teleost *Platypoecilus maculatus. J. Morph.* **52**, 115–53.

Yagamazaki, F. & Donaldson, E. M. (1968). The spermiation of goldfish (*Carassius auratus*) as a bioassay for salmon (*Oncorhynchus tshawitscha*) spermiation. *Gen. Comp. Endocrin.* **10**, 383–91.

Amphibians

Brökelmann, J. (1964). Über die Stütz- und Zwischenzellen des Froschhodens während des spermatogenetischen Zyklus. *Z. Zellforsch. mikrosk. Anat.* **64**, 429–61.

Burgos, M. H. & Vitale-Calpe, R. (1967). The mechanism of spermiation in the toad. *Am. J. Anat.* **120**, 227–52.

Bustos, E. & Cubillos, M. (1967). Ciclo celular en la e spermatogénesis de *Bufo spinolosus* Wiegman. Estudio radioautográfico preliminar. *Biologica* **40**, 62–8.

Bustos-Obregón, E. & Alliende, C. (1973). Spermatogonial renewal in amphibian testis. I. Mitotic activity in the breeding season. *Arch. Biol., Bruxelles* **84**, 329–39.

Bustos-Obregón, E., Alliende, C. & Schmiede, P. (1973). Spermatogonial renewal in amphibian testis. II. Seasonal variations and effect of temperature. *Arch. Biol., Bruxelles* **84**, 465–74.

de Robertis, D. E. P., Burgos, M. H. & Breyter, E. (1946). Action of anterior pituitary on Sertoli cells and the release of toad spermatozoa. *Proc. Soc. exp. Biol. Med.* **61**, 20–2.

Dongen, W. J. van, Ballieux, R. E. & Geursen, H. J. (1960). Spermiation in the common frog (*Rana temporaria*). III. Histochemical and chemical investigations. *Proc. K. ned. Akad. Wet.* **63**, 257–63.

Flemming, W. (1887). Neue Beiträge zur Kenntnis der Zelle. I. Die Kerntheilung bei den Spermatocyten von *Salamandra maculosa*. *Arch. mikrosk. Anat.* **29**, 389–463.

Gurwitsch, A. (1911). Untersuchungen über den zeitlichen Faktor der Zellteilungen. II. Mitteilung über das Wesen und das Vorkommen der Determination der Zellteilungen. *Arch. Entw.Mech.* **32**, 447–71.

Humphrey, R. R. (1921). The interstitial cells of the urodele testis. *Am. J. Anat.* **29**, 213–79.

Humphrey, R. R. (1922). The multiple testes in urodeles. *Biol. Bull.* **43**, 45–67.

Humphrey, R. R. (1925). A modification of the urodele testis resulting from germ cell degeneration. *Biol. Bull.* **48**, 145–65.

Lofts, B. (1964). Seasonal changes in the functional activity of the interstitial and spermatogenetic tissue of the green frog, *Rana esculenta*. *Gen. Comp. Endocrin.* **4**, 550–62.

Lofts, B. (1968). Patterns of testicular activity. In *Perspectives in Endocrinology*, ed. E. J. W. Barrington & C. B. Jørgensen, pp. 239–303. New York: Academic Press.

Lofts, B. & Boswell, C. (1960). Cyclical changes in the distribution of the testis lipids in the common frog, *Rana temporaria*. *Nature, Lond.* **187**, 708–9.

Meves, F. (1896). Ueber die Entwicklung der männlichen Geschlechtszellen von *Salamandra maculosa*. *Arch. mikrosk. Anat.* **48**, 1–83.

Oordt, P. G. W. J. van (1956). *Regulation of the Spermatogenetic Cycle in the Common Frog* (*Rana temporaria*). Arnheim: G. W. van der Wiel.

Oordt, P. G. W. J. van (1960). The influence of the internal and external factors in the regulation of the spermatogenetic cycle in amphibia. *Symp. Zool. Soc., Lond.* **2**, 29–52.

Poska-Teiss, L. (1933). Spermatogonien von *Bufo vulgaris* L. und ihr Vergleich mit larvalen somatischen Zellen desselben Tieres. *Z. Zellforsch. mikrosk. Anat.* **17**, 347–419.

Sentein, P. & Temple, D. (1971). Développement d'un complexe réticulaire au cours de l'évolution des spermatogonies primaries de *Triturus helveticus* Razoumowsky. *Arch. Anat. microsc. Morphol. exp.* **60**, 83–94.

Stieve, H. (1920). Die Entwicklung der Keimzellen des Grottenolms (*Proteus anguineus*). I Teil. Die Spermatocytogenese. *Arch. mikrosk. Anat.* **93**, 141–313.

Vallette St George, von la (1878). Über die Genese der Samenkörper. IV. Die Spermatogenese bei den Amphibien. *Arch. mikrosk. Anat.* **12**, 797–825.

Witschi, E. (1914). Studien über die Geschlechtsbestimmung bei den Fröschen. *Arch. mikrosk. Anat.* **86**, 1–50.

Witschi, E. (1924). Die Entwicklung der Keimzellen der *Rana temporaria*. *Z. Zellforsch. mikrosk. Anat.* **1**, 523–61.

Amniotes

Barr, A. B. (1973). Timing of spermatogenesis in four non-human primate species. *Fertil. Steril.* **24**, 281–9.

Branca, A. (1924). Les canalicules testiculaires et la spermatogénèse de l'Homme. *Arch. Zool. exp. gén.* **62**, 53–252.

Brendston, W. E. & Desjardin, C. (1974). The cycle of the seminiferous epithelium and spermatogenesis in the bovine testis. *Am. J. Anat.* **140**, 167–80.

Bustos-Obregón, E. (1974). Description of the boundary tissue of human seminiferous tubules under normal and pathological conditions. *Verh. anat. Ges.* **68**, 197–201.

Bustos-Obregón, E., Courot, M., Fléchon, J. E., Hochereau-de Reviers, M. T. & Holstein, A. F. (1975). Morphological appraisal of gametogenesis. Spermatogenetic process in mammals with particular reference to man. *Androl.* **7**, 141–62.

Bustos-Obregón, E. & Holstein, A. F. (1973). On structural patterns of the lamina propria of human seminiferous tubules. *Z. Zellforsch. mikrosk. Anat.* **141**, 413–25.

Cavazos, L. F. (1951). Spermatogenesis of the horned lizard *Phrynosoma cornutum*. *Am. Nat.* **85**, 373–9.

Clark, R. V. (1976). Three dimensional organization of testicular interstitial tissue and lymphatic space. *Anat. Rec.* **184**, 203–26.

Cleland, K. W. (1951). The spermatocytic cycle of the guinea pig. *Aust. J. Sci. Res. Ser. B* **4**, 344–69.

Clermont, Y. (1954). Cycle de l'épithélium séminal et mode de renouvellement des spermatogonies chez le Hamster. *Rev. Can. Biol.* **3**, 208–45.

Clermont, Y. (1958). Structure de l'épithélium séminal et mode de renouvellement des spermatogonies chez le canard. *Arch. Anat. microsc.* **47**, 47–66.

Clermont, Y. (1960). Cycle of the seminiferous epithelium of the guinea pig. A method for identification of the stages. *Fertil. Steril.* **11**, 563–73.

Clermont, Y. (1962). Quantitative analysis of spermatogenesis of the rat: a revised model for the renewal of spermatogonia. *Am. J. Anat.* **111**, 111–29.

Clermont, Y. (1963). The cycle of seminiferous epithelium in man. *Am. J. Anat.* **112**, 35–45.

Clermont, Y. (1967). Cinétique de la spermatogenèse chez les Mammifères. *Arch. Anat. microsc.* **56** (Suppl.), 7–60.

Clermont, Y. (1972). Kinetics of spermatogenesis in mammals: seminiferous epithelium cycle and spermatogonial renewal. *Physiol. Rev.* **52**, 198–236.

Clermont, Y. & Antar, M. (1973). Duration of the cycle of the seminiferous epithelium and the spermatogonial renewal in the monkey *Macaca arctoides*. *Am. J. Anat.* **136**, 153–65.

Clermont, Y. & Bustos-Obregón, E. (1968). Re-examination of spermatogonial renewal in the rat by means of seminiferous tubules mounted 'in toto'. *Am. J. Anat.* **122**, 237–48.

Clermont, Y. & Harvey, S. C. (1965). Duration of the cycle of the seminiferous epithelium of normal, hypophysectomized and hypophysectomized-hormone treated albino rats. *Endocrin.* **76**, 80–9.

Clermont, Y. & Perey, B. (1957). The stage of the cycle of the seminiferous epithelium of the rat: practical definitions in PA–Schiff–hematoxylin and hematoxylin–eosin stained sections. *Rev. Can. Biol.* **16**, 451–526.

Clermont, Y. & Trott, M. (1969). Duration of the cycle of the seminiferous epithelium in the mouse and hamster determined by means of H^3-thymidine and radioautography. *Fertil. Steril.* **20**, 805–17.

Courot, M., Hocherau-De Reviers, M. T. & Ortavant, R. (1970). Spermatogenesis. In *The Testis*, ed. A. D. Johnson, W. R. Gomes & N. L. Vandemark, vol. 1, chap. 6, pp. 339–432. New York: Academic Press.

Dalcq, A. (1921). Etude de la spermatogenèse chez l'orvet (*Anguis fragilis* Linn.). *Arch. Biol., Paris* **31**, 347–452.

Dalcq, A. (1973). Cyto-morphologie normale du testicule et spermatogenèse chez les mammifères. *Mém. Acad. r. Méd. Belg.* **46**, 1–834.

Davis, J. R., Langford, G. A. & Kirby, P. J. (1970). The testicular capsule. In *The Testis*, ed. A. D. Johnson, W. R. Gomes & N. L. Vandemark, vol. 1, pp. 281–337. New York: Academic Press.

De Kretser, D. M. (1969). Ultrastructural features of human spermiogenesis. *Z. Zellforsch. mikrosk. Anat.* **98**, 477–505.

De Rooij, D. G. (1968). Stem cell renewal and duration of spermatogonial cycle in the gold hamster. *Z. Zellforsch. mikrosk. Anat.* **89**, 133–6.

De Rooij, D. G. (1973). Spermatogonial stem cell renewal in the mouse. I. Normal situation. *Cell Tiss. Kinet.* **6**, 281–97.

Disselhorst, R. (1908). Gewichts- und Volumen-zunahme der männlichen Keimdrüse bei Vögeln und Säugern in der Paarungszeit: Unabhängigkeit des Wachstums. *Anat. Anz.* **32**, 113–17.

Dym, M. & Fawcett, D. W. (1970). The blood–testis barrier in the rat and the physiological compartmentation of the seminiferous epithelium. *Biol. Reprod.* **3**, 308–26.

Farner, D. S. (1975). Photoperiodic controls in the secretion of gonadotropins in birds. *Am. Zool.* **15** (Suppl. 1), 117–35.

Fawcett, D. W. (1975a). Ultrastructure and function of the Sertoli cell. In *Handbook of Physiology Section 7*, vol. 5, *Male Reproductive System*, ed. D. W. Hamilton & R. O. Greep, chap. 2, pp. 21–55. Washington, DC: American Physiological Society.

Fawcett, D. W. (1975b). The mammalian spermatozoon. *Dev. Biol.* **44**, 394–436.

Fawcett, D. W., Heidger, P. M. & Leak, L. V. (1969). Lymph vascular system of the interstitial tissue of the testis as revealed by electron microscopy. *J. Reprod. Fertil.* **19**, 109–19.

Fawcett, D. W., Neaves, W. & Flores, M. N. (1973). Comparative observations on intertubular lymphatics and the organization of the interstitial tissue of the mammalian testis. *Biol. Reprod.* **9**, 500–32.

Foote, R. H., Swierstra, E. E. & Hunt, W. L. (1972). Spermatogenesis in the dog. *Anat. Rec.* **173**, 341–52.

Fritz, I. B. (1973). Selected topics on the biochemistry of spermatogenesis. *Curr. Top. Cell Regul.* **7**, 129–74.

Gunn, S. A. & Gould, R. C. (1975). Vasculature of the testis and adnexa. In *Handbook of Physiology Section 7*, vol. 5, *Male Reproductive System*, ed. D. W. Hamilton & R. O. Greep, chap. 5, pp. 117–42. Washington, DC: American Physiological Society.

Hedinger, C. & Weber, E. (1973). Zur Struktur des menschlichen Hodens. *Z. Anat. Entw. Gesch.* **139**, 217–26.

Heller, C. G. & Clermont, Y. (1964). Kinetics of the germinal epithelium in man. *Recent Prog. Horm. Res.* **20**, 545–75.

Herland, M. (1933). Recherches histologiques et expérimental sur les variations cycliques du testicule et des caractères sexuels secondaires chez les Reptiles. *Arch. Biol.* **44**, 347–468.

Hilscher, B., Hilscher, W., Bülthoff-Ohnolz, B., Krämer, U., Birke, A., Pelzer, H. & Gauss, G. (1974). Kinetics of gametogenesis. I. Comparative histological and autoradiographic studies of oocytes and transitional prospermatogonia during oogenesis and prespermatogenesis. *Cell Tiss. Res.* **154**, 443–70.

Hilscher, W. & Makoski, H. (1968). Histologische und autoradiographische Untersuchungen zur 'Präspermatogenese' und 'Spermatogenese' der Ratte. *Z. Zellforsch. mikrosk. Anat.* **86**, 327–50.

Hochereau-de Reviers, M. T. (1970). Etudes des divisions spermatogoniales et du renouvellement de la spermatogonie-souche chez le toureau. Thesis, I.N.R.A.–C.R.V.Z. Lab. physiol. reproduct. Nouzilly, France.

Holstein, A. F. & Wartenberg, H. (1970). On the cyto-morphology of human spermatogenesis. *Adv. Androl.* **1**, 8–12.

Huckins, C. (1971*a*). The spermatogonial stem cell population in adult rats. I. Their morphology, proliferation and maturation. *Anat. Rec.* **169**, 533–58.

Huckins, C. (1971*b*). The spermatogonial stem cell population in adult rats. II. A radioautographic analysis of their cell cycle properties. *Cell Tiss. Kinet.* **4**, 313–34.

Huckins, C. (1971*c*). The spermatogonial stem cell population in adult rats. III. Evidence for a long-cycling population. *Cell Tiss. Kinet.* **4**, 335–49.

Johnson, F. P. (1934). Dissections of human seminiferous tubules. *Anat. Rec.* **59**, 187–99.

Johnson, O. W. & Buss, I. O. (1967). The testis of the African elephant (*Loxodonta africans*). I. Histological features. *J. Reprod. Fertil.* **13**, 11–21, 23–30.

Kenelly, J. J. (1972). Coyote reproduction. I. The duration of the spermatogenic cycle and epididymal sperm transport. *J. Reprod. Fertil.* **31**, 163–70.

Kirillow, S. (1912). Die Spermiogenese beim Pferde. *Arch. mikrosk. Anat.* **79**, 125–47.

Kramer, M. F. (1960). Spermatogenesis big de stier. Thesis, Utrecht.

Leblond, C. P. & Clermont, Y. (1952*a*). Definition of the stages of the cycle of the seminiferous epithelium of the rat. *Ann. N.Y. Acad. Sci.* **55**, 548–84.

Leblond, C. P. & Clermont, Y. (1952*b*). Spermiogenesis of rat, mouse, hamster and guinea pig as revealed by the 'periodic acid-fuchsin sulfurous acid' technique. *Am. J. Anat.* **90**, 167–216.

Lofts, B. (1968). Patterns of testicular activity. In *Perspectives in Endocrinology*, ed. E. J. W. Barrington & C. B. Jørgensen, chap. 4, pp. 239–304. New York: Academic Press.

Lofts, B. (1969). Seasonal cycles in reptilian testes. *Gen. Comp. Endrocrinol., Suppl. 2*, 147–55.

Moens, P. B. & Hugenholtz, A. D. (1975). The arrangement of germ cells in the rat seminiferous tubule: an electronmicroscopic study. *J. Cell Sci.* **19**, 487–507.

Monesi, V. (1962). Autoradiographic study of DNA synthesis and the cell cycle in the spermatogonia and spermatocytes in the mouse testis using tritiated thymidine. *J. Cell Biol.* **14**, 1–18.

Monesi, V. (1965). Synthetic activities during spermatogenesis in the mouse: RNA and protein. *Exp. Cell Res.* **39**, 197–224.

Oakberg, E. F. (1956). A description of spermatogenesis in the mouse and its use in analysis of the cycle of the seminiferous epithelium and germ cell renewal. *Am. J. Anat.* **99**, 507–16.

Oakberg, E. F. (1971). Spermatogonial stem cell renewal in the mouse. *Anat. Rec.* **169**, 515–32.

Ortavant, R. (1958). Le cycle spermatogénétique chez le bélier. PhD Thesis, Paris.

Ozon, R. (1972). Androgens in fishes, amphibians, reptiles and birds. In *Steroids in Non-mammalian Vertebrates*, ed. D. R. Idler, pp. 329–89. New York: Academic Press.

Perey, B., Clermont, Y. & Leblond, C. P. (1961). The wave of the seminiferous epithelium in the rat. *Am. J. Anat.* **108**, 47–77.

Regaud, C. (1901). Etudes sur la structure des tubes séminifères et sur la spermatogénèse chez les Mammifères. *Arch. Anat. microsc.* **4**, 101–56, 231–80.

Risley, P. L. (1938). Seasonal changes in the testis of the musk turtle, *Sternotherus adoratus* L. *J. Morph.* **63**, 301–17.

Roosen-Runge, E. C. (1961). The rete testis in the albino rat: its structure, development and significance. *Acta anat.* **45**, 1–30.

Roosen-Runge, E. C. (1962). The process of spermatogenesis in mammals. *Biol. Rev.* **37**, 343–77.

Roosen-Runge, E. C. & Giesel, L. O. (1950). Quantitative studies on spermatogenesis in the albino rat. *Am. J. Anat.* **87**, 1–30.

Ross, M. H. (1967). The fine structure and development of the peritubular contractile cell component in the seminiferous tubules of the mouse. *Am. J. Anat.* **121**, 523–58.

Rowley, M., Berlin, J. D. & Heller, C. G. (1971). The ultrastructure of four types of human spermatogonia. *Z. Zellforsch. mikrosk. Anat.* **112**, 139–57.

Rowley, M. & Heller, C. (1971). Quantitation of the cells of the seminiferous epithelium of the human testis employing the Sertoli cell as a constant. *Z. Zellforsch. mikrosk. Anat.* **115**, 461–72.

Schöneberg, K. (1913). Die Samenbildung bei den Enten. *Arch. mikrosk. Anat.* **83**, 324–69.

Setchell, B. P. (1967). The blood–testicular fluid barrier in sheep. *J. Physiol.* **189**, 63–5.

Setchell, B. P. (1970). Testicular blood supply, lymphatic drainage, and secretion of fluid. In *The Testis*, ed. A. D. Johnson, W. R. Gomes & N. L. Vandemark, vol. 1, chap. 3, pp. 101–239. New York: Academic Press.

Setchell, B. P. & Waites, G. M. H. (1975). The blood–testis barrier. In *Handbook of Physiology Section 7*, vol. 5, *Male Reproductive System*, ed. D. W. Hamilton & R. O. Greep, chap. 6, pp. 143–72. Washington, DC: American Physiological Society.

Steinberger, E. & Steinberger, A. (1974). Hormonal control of testicular function in mammals. In *Handbook of Physiology Section 7*, vol. 4, *The Pituitary Gland and Its Neuroendocrine Control*, ed. E. Knobil & W. H. Sawyer, chap. 34, pp. 325–45. Washington, DC: American Physiological Society.

Steinberger, E. & Steinberger, A. (1975). Spermatogenic function of the testis. In *Handbook of Physiology Section 7*, vol. 5, *Male Reproductive System*, ed. D. W. Hamilton & R. O. Greep, chap. 1, pp. 1–19. Washington, DC: American Physiological Society.

Stieve, H. (1930). Männliche Genitalorgane. In *Handbuch der Mikroskopischen Anatomie des Menschen*, ed. W. v. Möllendorf, vol. 7, part 2, pp. 1–387. Berlin: J. Springer.

Swierstra, E. E. (1968). Cytology and duration of the cycle of the seminiferous epithelium of the boar; duration of the spermatozoan transit through the epididymis. *Anat. Rec.* **161**, 171–86.

Swierstra, E. E. & Foote, R. H. (1963). Cytology and kinetics of spermatogenesis in the rabbit. *J. Reprod. Fertil.* **5**, 309–22.

Swierstra, E. E., Gebauer, M. R. & Pickett, B. W. (1974). Reproductive physiology of the stallion. I. Spermatogenesis and testis composition. *J. Reprod. Fertil.* **40**, 113–24.

Thibault, C. (1969). La spermatogenèse chez les mammifères. In *Traité de Zoologie*, ed. P. P. Grasse, pp. 716–98. Paris: Masson et Cie.

Tobias, P. V. (1956). *Chromosomes Sex-Cells and Evolution in a Mammal.* London: Percy Lund, Humphries & Co.

Unsicker, K. & Burnstock, G. (1975). Myoid cells in the peritubular tissue (Lamina propria) of the reptilian testis. *Cell Tiss. Res.* **163**, 545–60.

Wilhoft, D. C. & Reiter, E. O. (1965). Seasonal cycle of the lizard, *Leiolopisma fuscum*, a tropical Australian skink. *J. Morph.* **116**, 379–88.

Chapter 9. Spermatogonia and kinetics of spermatogenesis

Amann, R. P. (1970). Sperm production rates. In *The Testis*, ed. A. D. Johnson, W. R. Gomes & N. L. Vandemark, chap. 3. New York: Academic Press.

Bugnion, E. (1906). La signification des faisceaux spermatiques. *Bibl. anat.* **16**, 5–52.

Chandley, A. C. & Bateman, A. J. (1962). Timing of spermatogenesis in *Drosophila melanogaster* using tritiated thymidine. *Nature, Lond.* **193**, 299–300.

Clermont, Y. (1969). Two classes of spermatogonial stem cells in the monkey (*Cercopithecus aethiops*). *Am. J. Anat.* **126**, 57–72.

Clermont, Y. & Bustos-Obregón, E. (1968). Re-examination of spermatogonial renewal in the rat by means of seminiferous tubules mounted 'in toto'. *Am. J. Anat.* **122**, 237–48.

Clermont, Y. & Leblond, C. P. (1959). Differentiation and renewal of spermatogonia in the monkey *Macacus rhesus. Amer. J. Anat.* **104**, 237–72.

Clermont, Y., Leblond, C. P. & Messier, B. (1959). Durée du cycle de l'épithélium séminal du Rat. *Arch. Anat. microsc. Morph. exp.* **48**, 37–56.

Dym, M. & Fawcett, D. (1971). Further observations on the numbers of spermatogonia, spermatocytes, and spermatids connected by intercellular bridges in the mammalian testis. *Biol. Reprod.* **4**, 195–215.

Edwards, J. S. (1969). On the reproduction of *Prionoplus reticularis* (Coleoptera, Cerambycidae) with general remarks on reproduction in the Cerambycidae. *Q. J. microsc. Sci.* **102**, 519–29.

Hancock, J. L. (1972). Spermatogenesis and sperm defects. In *The Genetics of the Spermatozoon, Edinburgh Symposium on the Genetics of the Spermatozoon*, ed. R. A. Beatty & S. Gluecksohn-Waelsch, pp. 121–30. Edinburgh: Beatty & Gluecksohn-Waelsch.

Hilscher, B., Hilscher, W. & Maurer, W. (1966). Autoradiographische Bestimmung der Generationszeiten und Teilphasen der verschiedenen Spermatogonien-Generationen der Ratte. *Naturwissenschaften* **53**, 415–16.

Huckins, C. (1971a). The spermatogonial stem cell population in adult rats. I. Their morphology, proliferation and maturation. *Anat. Rec.* **169**, 533–58.

Huckins, C. (1971b). The spermatogonial stem cell population in adult rats. II. A radioautographic analysis of their cell cycle properties. *Cell Tiss. Kinet.* **4**, 313–34.

Huckins, C. (1971c). The spermatogonial stem cell population in adult rats. III. Evidence for a long-cycling population. *Cell Tiss. Kinet.* **4**, 335–49.

Lentz, T. L. (1966). *The Cell Biology of Hydra*. Amsterdam: North-Holland.

Oakberg, E. F. (1971). Spermatogonial stem cell renewal in the mouse. *Anat. Rec.* **169**, 515–32.

Lévi, C. (1956). Etude des Halisarca des Roscoff. Embryologie et systématiques des Demosponges. *Arch. Zool. exp. gén.* **93**, 1–184.

Roosen-Runge, E. C. (1955). Untersuchungen über die Degeneration samenbildender Zellen in der normalen Spermatogenese der Ratte. *Z. Zellforsch. mikrosk. Anat.* **41**, 221–35.

Roosen-Runge, E. C. (1960). Cyclic spermatogenesis in a hydromedusa (*Phialidium* L.) as compared with the rat. *Anat. Rec.* **136**, 267 (abstract).

Roosen-Runge, E. C. (1962). Behavior of isolated male and female gonads of the hydromedusa *Phialidium gregarium* (L. Agassiz). *Anat. Rec.* **142**, 273 (abstract).

Roosen-Runge, E. C. & Barlow, F. D. (1953). Quantitative studies on human spermatogenesis. I. Spermatogonia. *Am. J. Anat.* **93**, 143–69.

Sado, T. (1961). Spermatogenesis of the silkworm and its bearing on the radiation induced sterility. *Jap. J. Genet., Suppl.* **36**, 136–51.

Schreiber, G. (1949). Statistical and physiological studies on the interphasic growth of nuclei. *Biol. Bull.* **97**, 187–205.

Tihen, J. A. (1946). An estimate of the number of cell generations preceding sperm formation in *Drosophila melanogaster. Am. Nat.* **80**, 389–92.

Chapter 10. Degeneration; polymorphism; genetic control

Barr, A. E., Moore, D. J. & Paulsen, C. A. (1971). Germinal cell loss during human spermatogenesis. *J. Reprod. Fertil.* **25**, 75–80.

Beatty, R. A. (1972). The genetics of size and shape of spermatozoan organelles. In *The Genetics of the Spermatozoon, Edinburgh Symposium on the Genetics of the Spermatozoon*, ed. R. A. Beatty & S. Gluecksohn-Waelsch, pp. 97–115. Edinburgh: Beatty & Gluecksohn-Waelsch.

Beatty, R. A. (1970). The genetics of the mammalian gamete. *Biol. Rev.* **45**, 73–120.

Beatty, R. A. & Sidhu, N. S. (1969). Polymegaly of spermatozoan length and its genetic control in *Drosophila* species. *Proc. R. Soc. Edinburgh, Ser. B* **71**, 14–28.

Czernosvitov, L. (1931). Studien über die Spermaresorption. III. Die Samenresorption bei den Tricladen. *Zool. Jahrb. (Anat.)* **54**, 295–330.

Champy, C. (1913). Recherches sur la Spermatogénèse des Batraciens. *Arch. Zool. exp. gén.* **52**, 13.

Fawcett, D. W. (1972). Observations on cell differentiation and organelle continuity in spermatogenesis. In *The Genetics of the Spermatozoon, Edinburgh Symposium on the Genetics of the Spermatozoon*, ed. R. A. Beatty & S. Gluecksohn-Waelsch, pp. 37–68. Edinburgh: Beatty & Gluecksohn-Waelsch.

Hendelberg, J. (1969). On the development of different types of spermatozoa from spermatids with two flagella in the *Turbellaria* with remarks on the ultrastructure of the flagella. *Zool. Bidr., Uppsala* **38**, 1–50.

Holstein, A. F. (1975). Morphologische Studien an abnormen Spermatiden und Spermatozoen des Menschen. *Virchow's Arch. path. Anat. Physiol.* **367**, 93–112.

Hyman, L. (1951). *The Invertebrates*, vol. 3, *Acantocephala, Aschelminthes and Entoprocta*, p. 147. New York: McGraw-Hill.

MacLeod, J. (1970). The significance of deviations in human sperm morphology. *Adv. exp. Med. Biol.* **10**, 481–94.

Meves, F. (1903). Über oligopyrene und apyrene Spermien und ihre Enstehung nach Beobachtungen an *Paludina* und *Apygaera*. *Arch. mikrosk. Anat. Entw.-Mech.* **61**, 1–84.

Raven, C. P. (1961). *Oogenesis: The Storage of Developmental Information. International Series Monograph in Pure and Applied Biology, Division Zoology*, vol. 10. Oxford: Pergamon Press.

Roosen-Runge, E. C. (1955). Untersuchungen über die Degeneration samenbildender Zellen in der normalen Spermatogenese der Ratte. *Z. Zellforsch. mikrosk. Anat.* **41**, 221–35.

Roosen-Runge, E. C. (1973). Germinal-cell loss in normal metazoan spermatogenesis. *J. Reprod. Fertil.* **35**, 339–48.

Saunders, J. W. & Fallon, J. F. (1966). Cell death in morphogenesis. In *Twenty-fifth Symposium of the Society of Developmental Biology*, ed. M. Locke, pp. 289–314. New York: Academic Press.

Schleip, W. (1912). Das Verhalten des Chromatins bei *Angiostomum* (Rhabelanema) *nigrovenosum*. *Arch. Zellforsch.* **7**, 87–138.

Skakkebaek, N. E., Bryant, J. I. & Philip, J. (1973). Studies on meiotic chromosomes in infertile men and controls with normal karyotypes. *J. Reprod. Fertil.* **35**, 23–36.

Strogonova, N. S. (1963). Characteristics of spermatogenesis in some bivalve molluscs. *Vest. Mosk. Univ. Ser. 6, Biol. Pochvovedenie* **18**, 25–34. *Biol. Abstr.* **46**, no. 69027.

Tretjakoff, D. (1905). Die Spermatogenese bei *Ascaris megalocephala*. *Arch. mikrosk. Anat.* **65**, 383–438.

Chapter 11. Microenvironment of spermatogenesis. Compartments and auxiliary cells

Fawcett, D. W. (1975). Ultrastructure and function of the Sertoli cell. In *Handbook of Physiology Section 7*, vol. 5, *Male Reproductive System*, ed. D. W. Hamilton & R. O. Greep, chap. 2, pp. 21–55. Washington, DC: American Physiological Society.

Hansson, V., Ritzén, E. M., French, F. S. & Nayfeh, S. N. (1975). Androgen transport and receptor mechanisms in testis and epididymis. In *Handbook of Physiology Section 7*, vol. 5, *Male Reproductive System*, ed. D. W. Hamilton & R. O. Greep, pp. 173–201. Washington, DC: American Physiological Society.

Raven, C. P. (1961). *Oogenesis: the Storage of Developmental Information*. Oxford: Pergamon Press.

Setchell, B. P. & Waites, G. M. H. (1975). The blood–testis barrier. In *Handbook of Physiology Section 7*, vol. 5, *Male Reproductive System*, ed. D. W. Hamilton & R. O. Greep, chap. 6, pp. 143–72. Washington, DC: American Physiological Society.

Woolley, P. (1975). The seminiferous tubules in dasyurid marsupials. *J. Reprod. Fertil.* **45**, 255–61.

Chapter 12. Conclusions

Bé, A. W. & Anderson, O. R. (1976). Gametogenesis in planktonic Foraminifera. *Science*, **192**, 890–2.

Glossary

Ancel, P. & Bouin, P. (1926). Recherches expérimentales sur l'origine des gonocytes dans le testicule des Mammifères. *C. R. Assoc. Anat.* **21**, 1–11.

André, J. & Rouiller, C. (1957). L'ultrastructure de la membrane nucléaire des ovocytes de l'araignée (*Tegenaria domestica* Clark). In *Proceedings of the European Conference on Electron Microscopy*, Stockholm, 1956, pp 162–4. New York: Academic Press.

Ankel, W. E. (1930). Die atypische Spermatogenese von *Janthina* (Prosobranchia, Ptenoglossia). *Z. Zellforsch. mikrosk. Anat.* **11**, 491–608.

Baer, C. E. von (1827). Beiträge zur Kenntnis der niederen Thiere. *Nova Acta Leop. Carol.* **13**, 640.

Ballowitz, E. (1899). Untersuchungen über die Struktur der Spermatozoen, zugleich ein Beitrag zur Lehre vom feineren Bau der kontraktilen Elemente. Die Spermatozoen der Insekten. *Z. wiss. Zool.* **50**, 383–6.

Beddard, F. E. (1895). *A Monograph on the Order of Oligochaeta*. Oxford: Clarendon Press.

Bishop, M. W. H. & Walton, A. (1960). Spermatogenesis and the structure of the mammalian spermatozoa. In Marshall's *Physiology of Reproduction*, ed. A. S. Parkes, vol. 1, pt. 2. Harlow: Longman.

D'Ancona, V. (1943). Nuove recirche sulla determinazione sessuale dell'anguilla. *Arch. Oceanogr. Limnol., Roma* **3**, 159–271.

Ebner, V. von (1888). Zur Spermatogenese bei den Säugetieren. *Arch. mikrosk. Anat.* **31**, 236–92.

Ebner, V. von (1902). Männliche Geschlechtsorgane. In A. Koelliker's *Handbuch der Gewebelehre des Menschen*, vol. 3, pp. 401–505. Leipzig: Engelmann.

Eddy, E. M. (1975). Germ plasm and the differentiation of the germ cell line. *Int. Rev. Cytol.* **43**, 229–80.

Farmer, J. B. & Moore, J. E. S. (1905). On the maiotic phase (reduction divisions) in animals and plants. *Q. J. microsc. Sci.* **48**, 489–558.

Felix, W. & Bühler, A. (1906). Die Entwicklung der Keimdrüsen und ihrer Ausführungsgänge. In *Entwicklungslehre der Wirbeltiere*, vol. 5, chap. 2, pp. 619–896. Jena: G. Fischer.

Gilson, G. (1886). Etude comparée de la spermatogénèse chez les arthropodes. *Cellule* **2**, 83–239.

Grier, H. J. (1975). Aspects of germinal cyst and sperm development in *Poecilia latipinna* (Teleostei: Poeciliidae). *J. Morph.* **146**, 229–50.

Grünberg, K. (1903). Untersuchungen über die Keim- und Nährzellen in den Hoden und Ovarien der Lepidopteren. *Z. wiss. Zool.* **74**, 327–95.

Hanisch, J. (1970). Die Blastostyle- und Spermienentwicklung von *Eudendrium racemosum* Cavolini. *Zool. Jahrb. (Anat.)* **87**, 1–62.

Hannah-Alava, A. (1965). The premeiotic stages of spermatogenesis. *Adv. Genet.* **13**, 157–226.

Hamilton, W. J., Boyd, J. D. & Mossman, H. W. (1957). *Textbook of Human Embryology*, 4th edn. Cambridge: W. Heffer & Sons.

Hardisty, M. W. (1965). Sex differentiation and gonadogenesis in lampreys. I. The ammo-coete gonads of the brook lamprey, *Lampetra planeri. J. Zool.* **146**, 305–45.

Hilscher, B., Hilscher, W., Bülthoff-Ohnolz, B., Krämer, U., Birke, A., Pelzer, H. & Gauss, G. (1974). Kinetics of gametogenesis. I. Comparative histological and autoradiographic studies of oocytes and transitional prospermatogonia during oogenesis and prespermatogenesis. *Cell Tiss. Res.* **154**, 443–70.

Hilscher, W. (1970). Kinetics of prespermatogenesis and spermatogenesis of the Wistar rat under normal and pathological conditions. *Morph. Aspects Androl.* **1**, 17–20.

Hilscher, W. & Makoski, H. (1968). Histologische und autoradiographische Untersuchungen zur 'Präspermatogenese' und 'Spermatogenese' der Ratte. *Z. Zellforsch. mikrosk. Anat.* **86**, 327–50.

Hoage, T. R. & Kessel, R. G. (1968). An electron microscope study of the process of differentiation during spermatogenesis in the drone honey bee (*Apis mellifera* L.) with special reference to centriole replication and elimination. *J. Ultrastruct. Res.* **24**, 6–32.

Hyman, L. (1951). Acantocephala, Aschelminthes and Entoprocta. In *The Invertebrates*, vol. 3, p. 245. New York: McGraw-Hill.

Leblond, C. P. & Clermont, Y. (1952). Definition of the stages of the cycle of the seminiferous epithelium in the rat. *Ann. N. Y. Acad. Sci.* **55**, 548–73.

Marshall, A. J. & Lofts, B. (1956). The Leydig cell homologue in certain teleost fishes. *Nature, Lond.* **177**, 705–6.

Meves, F. (1903). Über oligopyrene und apyrene Spermien und ihre Entstehung nach Beobachtungen an Paludina and Pygaera. *Arch. mikrosk. Anat. Entw.-Mech.* **61**, 1.

Oppermann, E. (1935). Zur Entstehung der Riesenspermien von *Argas columbarum* (Shaw) (reflexus F.). *Z. mikrosk.-anat. Forsch.* **37**, 538–60.

Oordt, G. J. van & Klomp, H. (1946). Effects of oestrone and gonadotrophin administration in the male toad. *Proc. K. ned. Akad. Wet. (Sect. Sci.)* **49**, 565–70.

Regaud, C. (1901). Etudes sur la structure des tubes séminifères et sur la spermatogénèse chez les Mammifères. *Arch. Anat. microsc.* **4**, 101–56, 231–80.

Roosen-Runge, E. C. (1962). The process of spermatogenesis in mammals. *Biol. Rev.* **37**, 343–77.

Ruby, S. M. & McMillan, D. B. (1975). The interstitial origin of germinal cells in the testis of the stickleback *Culaea inconstans. J. Morph.* **145**, 295–318.

Sertoli, E. (1865). Dell'esistenza di particolari cellule ramificati nei canalicoli seminiferi del testiculo humano. *Morgagni* **7**, 31–9.

Stanley, H. P. (1966). The structure and development of the seminiferous follicle in *Scyliorhinus caniculus* and *Torpedo marmorata* (Elasmobranchii). *Z. Zellforsch. mikrosk. Anat.* **75**, 453–68.

Stephenson, J. (1930). *The Oligochaeta.* Oxford: Clarendon Press.

Szöllösi, A. (1975). Electron microscope study of spermiogenesis in *Locusta migratoria* (Insect orthoptera). *J. Ultrastruct. Res.* **50**, 322–46.

Tönniges, C. (1902). Beiträge zur Spermatogenese und Oogenese der Myriopoden. *Z. wiss. Zool.* **71**, 328–58.

Turner, C. L. (1919). The seasonal cycle in the spermary of the perch. *J. Morph.* **32**, 681–711.

Vallette St George, von la (1876). Ueber die Genese der Samenkörper. *Arch. mikrosk. Anat.* **12**, 797–807, 812–21.

Waldeyer, W. (1871). Eierstock und Nebeneierstock. In J. Stricker's *Handbuch der Gewebelehre*, p. 576. Leipzig: Engelmann.

Wilson, E. B. (1925). *The Cell in Development and Heredity.* New York: Macmillan.

Witschi, E. (1924). Die Entwicklung der Keimzellen von *Rana temporaria. Z. Zellforsch. mikrosk. Anat.* **1**, 523–61.

Word, B. H. Jr & Hobbs, H. H. Jr (1958). Observations on the testis of the crayfish *Cambarus montanus acuminatus* Faxon. *Trans. Am. Microsc. Soc.* **77**, 435–50.

Author index

Page numbers in bold type refer to illustrations.

Index of animal names

Page numbers in bold type refer to illustrations.

Subject index

Terms in italics are contained in the glossary. Page numbers in bold type refer to illustrations. All animal names are to be found in the index of animal names.

acinus, 58, 60
ammocoete larva, 103
amoebocyte, 23
ampulla, 103, 104, 116, 164
andrology, 20
apical cell, 77–82, 164–5
apyrene, **148**, **149**, 165
auxocyte, 8, 165
axial cell, 83

basal cell, 43–4, 46
basal lamina, 165
basement membrane, 53, 165
birth control, 20
bivalents, 7
blastostyle, *see* hydranth

cell, *see specific cell*
cell theory, 13, 14
chiasmata, 7
choanocyte, 24, 25
clone, *see* under spermatogonium
coelenchyma, 96
companion cell, 113, 165
covering cell, **23**
crossing over, 7
cycle: seasonal, 100, 109, 116, 129; *of seminiferous epithelium*, 83, 123ff., **125**, **128**, 165
cyst, **10**, **63**, **65**, 72–7, 111, 117, 155–6: boundary, 112; *definition*, 10, 105, 162, 165
cyst cell, 64: *definition*, 11, 113, 162, 165; degeneration, 114; development, **106**; differentiation, 74; phagocytosis, 74ff., 113
cytophore, 40, **56**, 84ff., **86**, 156, 162, 165, 167: degeneration, 145: of germ cells, 4, 32, 50, 73, 78, 117, 145–8; of nurse cells, 35, 40, 47

desmosomes, supporting cells, 113, **124**
deuterogonium, 103, 165
dimorphism, 37, 98, *see also* polymorphism
diplonema, 7, 165
diplotene, *see* diplonema
DNA synthesis, 6: in spermatogonia, 140

endocrine cell, 114
eupyrene, **148**, **149**, 165

follicle, 67, 84, 92, 103, 155, 165–6: formation, 24–5
follicle cell, 23, 49, 75, 113
follicular membrane (envelope), 24, 25
fusom, 70, *see also* intercellular bridges, syncytial connections

germ cell: degeneration, 4, 32, 50, 73, 78, 117, 145–8; differentiation, **38**
germ line, 31
germinal epithelium, 42, 46, 62, 166
glycogen: in cyst cells, 74, 119, 121
gonocyte, 4, 123, 166

hermaphroditism, 37, 38–9, 168
hydranth, 32

individualization process, 70
intercellular bridges, 10, 36, 70, 101, 134, *see also* syncytial connections
interstitial cell, 31, 76

'Kinoplasma Kugeln', 43

leptonema, 6, 166
leptotene, *see* leptonema
Leydig cell, 114, 116
limiting membrane, 123, 166
lobule, 107ff., 116, 155
lobule boundary cell, 112, 166